SCHOLASTIC
English & Maths Bumper Book Ages 5–6

2 books in 1

REVISION & PRACTICE

KS1 Year 1

Build knowledge and confidence in maths and English

SCHOLASTIC

First published in the UK by Scholastic, 2025

Scholastic Education, Bosworth Avenue, Warwick, CV34 6UQ

Scholastic Ireland, 89E Lagan Road, Dublin Industrial Estate, Glasnevin, Dublin, D11 HP5F

SCHOLASTIC and associated logos are trademarks and/or registered trademarks of Scholastic Inc.

© Scholastic, 2025

123456789 5678901234

A CIP catalogue record for this book is available from the British Library.

ISBN 978-0702-33631-7

Printed and bound in India by Replika Press Pvt. Ltd.

This product is made from material from well-managed, FSC®-certified forests and other controlled sources.

All rights reserved.

This book is sold subject to the condition that it shall not, by way of trade or otherwise, be lent, hired out or otherwise circulated in any form of binding or cover other than that in which it is published. No part of this publication may be reproduced, stored in a retrieval system, or transmitted in any form or by any other means (electronic, mechanical, photocopying, recording or otherwise) or used to train any artificial intelligence technologies without prior written permission of Scholastic Limited. Subject to EU law Scholastic Limited expressly reserves this work from the text and data mining exception.

Due to the nature of the web we cannot guarantee the content or links of any site mentioned.

We strongly recommend that teachers check websites before using them in the classroom.

Every effort has been made to trace copyright holders for the works reproduced in this book, and the Publishers apologise for any inadvertent omissions.

www.scholastic.co.uk

For safety or quality concerns:
UK: www.scholastic.co.uk/productinformation
EU: www.scholastic.ie/productinformation

Author
Catherine Casey

Editorial team
Rachel Morgan, Audrey Stokes, Christine Bruce, Julia Roberts, David & Jackie Link

Design team
Andrea Lewis, QBS Learning

Illustrations
Cover, p9, p63: Tigatelu/iStock

Robot: VIGE.CO/Shutterstock

QBS Learning: p30, p32 (flowers), p33, p38 (wink), p49 & 79 (bus), p54, p55 (house), p59, p60, p66, p70 (flowers), p71 (apples), p72, 80 & 98 (cubes), p73 (crayons), p74, p75 & 79 (dice), p77, p79 (bricks), p80, 82 & 86 (beads), p81 (dice), p81 (bird), p85, p90, p92 (pizza), p92 (eggs), p93 (flowers), p94 (cars), p95 (ladybirds), p99 (sweets), p100 (boy & girl), p100 (snakes), p102 (manual scales), p102 (digital scales), p103, p104, p105 (watering can), p105 (jugs), p106 (alarm clock), p109 (brush), p111 (boy), p114, p115

Other: p10 (rainbow) Tartila/Shutterstock; p11, 22 (zebra) d3verro/Shutterstock; p11, 19 (pizza) lawang design/Shutterstock; p12 (worm) VikiVector/iStock; p12 (nose) Maman Suryaman/Shutterstock; p13 (cactus) Elena Istomina/Shutterstock; p14 (rain cloud) Studio_G/Shutterstock; p14 (roller boots) Softachka/Shutterstock; p14 (tortoise) Simply Amazing/Shutterstock; p15 (snowman) Only _up/Shutterstock; p15 (broccoli) Magicleaf/Shutterstock; p15 (puzzle piece) sewonboy/Shutterstock; p16 (milkshake) aelitta/iStock; p17 (grasshopper) Nsit/Shutterstock; p17 (puddle) Sylfida/Shutterstock; p18, 29, 50 (sunflower) Dernkadel/Shutterstock; p18 (fox) d3verro/Shutterstock; p18 (plums) Andrii Bezvershenko/Shutterstock; p18 (snail) WinWin artlab/Shutterstock; p19, 28, 58 (owl) GoodStudio/Shutterstock; p19, 44 (frog) d3verro/Shutterstock; p19 (pterodactyl) Ianrward/Shutterstock; p19 (laptop) johavel/Shutterstock; p20 (sloth) Maquiladora/Shutterstock; p20, 56 (train) Tomacco/Shutterstock; p21 (sweets) Texturis/Shutterstock; p21 (lion) d3verro/Shutterstock; p21 (rocket) vivat/Shutterstock; p21 (hippo) d3verro/Shutterstock; p22, 27 (moon and stars) Begimail/Shutterstock; p22, 28 (girl jumping) Colorfuel Studio/Shutterstock; p23 (cheetah) d3verro/Shutterstock; p23 (dog) Good Studio/Shutterstock; p23, 44 (elephant) d3verro/Shutterstock; p24 (corn) Andrii Bezvershenko/Shutterstock; p24 (dinosaur & nest) Teguh Mujiono/Shutterstock; p24 (eggs) Mihai Gruianu/Shutterstock; p25 (bread) Neliakott/Shutterstock; p25 (snowman) Panda Vectors/Shutterstock; p25 (carrot) Andy.Illustrator/Shutterstock; p25, 48, 50, 83 (football) Kanate/Shutterstock; p26, 52 (cats) ayutaka/iStock; p27 (hats) Tartila/Shutterstock; p27 (tractor) Tanislava/Shutterstock; p27 (trees) vivat/Shutterstock; p27 (socks) RZN_desing/Shutterstock; p27, 58, 89, 96 (cupcakes) yuwnis07/Shutterstock; p27, 52 (boxes) olesia_g/Shutterstock; p27 (lamp) Kolonko/Shutterstock; p27 (paintbrushes) Dmytro K/Shutterstock; p27 (watches) StockSmartStart/Shutterstock; p27 (dresses) YummyBuum/Shutterstock; p28 (dishes) olly2polly/Shutterstock; p28 (tennis racket & ball) Handatko/Shutterstock; p29 (window) pikepicture/Shutterstock; p29 (door) Oxy_gen/Shutterstock; p29 (music notes) Kharom Pleedee/iStock; p32 (bird) spirkaart/Shutterstock; p32 (kitten) KoAlia_draws/Shutterstock; p32 (mouse) Mvshop/Shutterstock; p34, 39 (shoes) KutuzovaDesign/Shutterstock; p34 (paper) sumkinn/Shutterstock; p34 (boys) Tana-mame/iStock; p35 (zip) A 5/Shutterstock; p37 (clapping emojis) Sudowoodo/Shutterstock; p38, 46, 52 (paint splodge) Neliakott/Shutterstock; p38, 39 (woman thinking) mentalmind/Shutterstock; p38, 39 (sink) Fancy Tapis/Shutterstock; p38, 39, 105 (glass of water) Frogella/Shutterstock; p38, 39 (linked chains) Nadia Snopek/Shutterstock; p38 (shaking hands) LerinaInk/Shutterstock; p38, 39 (sinking boat) BlueRingMedia/Shutterstock; p38 (empty glass) Frogella/Shutterstock; p38, 39 (water tank) MaryValery/Shutterstock; p38, 39 (bank) Eloku/Shutterstock; p38 (horn) VectorPlotnikoff/Shutterstock; p40 (sandwich) LadadikArt/iStock; p40 (orange juice carton) shock77/iStock; p40 (dog with ball) YG Studio/Shutterstock; p40 (sign) Sensvector/Shutterstock; p40, 41 (chick & egg hatching) ivector/Shutterstock; p42 (van) Golden Sikorka/Shutterstock; p42 (vet) denayunebgt/Shutterstock; p42 (vest) johavel/Shutterstock; p43, 70 (glove) Look_Studio/Shutterstock; p43 (olive) lemono/iStock; p43 (heart) Graphics Studio MH. We offer you the best things/iStock; p43 (dove) elenabs/iStock; p44 (fish) d3verro/Shutterstock; p44 (fairy) Natalllenka.m/Shutterstock; p44 (feet) Julee Ashmead/Shutterstock; p44 (dolphin) d3verro/Shutterstock; p44 (alphabet) andreysharonov/Shutterstock; p44 (sphere) Sensvector/Shutterstock; p44, 110 (mobile phone) Carkhe/Shutterstock; p44 (trophy) CleanVector/Shutterstock; p46, 52 (yak) d3verro/Shutterstock; p46, 52 (yoghurt) Shany Muchnik/Shutterstock; p46, 47 (cherry) FoxGrafy/Shutterstock; p46, 47, 57 (bunny) mentalmind/Shutterstock; p46, 47 (woman) tynyuk/Shutterstock; p46, 47 (dog) Macrovector/Shutterstock; p46, 47 (celebration emoji) Tally18/Shutterstock; p46, 47 (berry) Ksenia Gromova/Shutterstock; p46, 47 (family) Tenstudio/Shutterstock; p48 (apples) Alody/Shutterstock; p50 (toothbrush) Anatolir/Shutterstock; p50 (snowball) MaryValery/Shutterstock; p50 (popcorn) Kumeko/Shutterstock; p50 (eyebrow) ivector/Shutterstock; p50 (ball) sabelskaya/iStock; p50 (hairbrush) lemono/iStock; p50 (flower) TheArtist/iStock; p50 (foot) Easy_Company/iStock; p50 (tooth) Jane_Kelly/iStock; p50 (sun) Davyd Kopych/iStock; p51 (raindrop) Net Vector/Shutterstock; p51 (snowflake) graficriver_icons_logo/Shutterstock; p51 (blueberries) Ksenia Gromova/Shutterstock; p51 (doormat) StockSmartStart/Shutterstock; p52 (cot) Quang Vinh Tran/Shutterstock; p52 (man) tynyuk/Shutterstock; p52 (girl) Galkin_K/iStock; p55 (robot) Nursery Art Design/Shutterstock; p55 (whale) d3verro/Shutterstock; p55 (snake) d3verro/Shutterstock; p57 (toy truck) SurfsUp/Shutterstock; p58 (magnifying glass) chuhastock/Shutterstock; p58 (boy) seamartini/iStock; p61 (slide) Bobrik74/Shutterstock; p64 (rockets) vivat/Shutterstock; p64 (planets) Ana Tivikova/Shutterstock; p67 (rosette) Valeriia Soloveva/Shutterstock; p68 (socks) Iryna_ZA/Shutterstock; p68 (bikes) Kuryanovich Tatsiana/Shutterstock; p69 (nests) Look_Studio/Shutterstock; p69 (eggs) Mihai Gruianu/Shutterstock; p72 (hands) joshya/Shutterstock; p75 (cupcake) sandraaa1994/Shutterstock; p76 (ice cream) ina9/Shutterstock; p76 (apples) Alody/Shutterstock; p80 (ladybird) Roman Kybus/Shutterstock; p80 (woodlice) pink.mousy/Shutterstock; p82 (clothes pegs) Angela Jones/Shutterstock; p87 (caterpillar) My-Sun-Shine/Shutterstock; p87 (beetle) WinWin artlab/Shutterstock; p89, 95 (strawberry) 4gladiator.studio44/Shutterstock; p91 (sweets) Texturis/Shutterstock; p91 (marble) BlueRingMedia/Shutterstock; p93, 100, 101, 108 (pencils) Naddya/Shutterstock; p94 (pasta) Sylfida/Shutterstock; p94 (butterfly) Fresh_Studio/Shutterstock; p95 (tables) Elegant Solution/Shutterstock; p95 (pile of pencils) Veronika_Decart/Shutterstock; p96 (buttons) ZK Fashion Design/Shutterstock; p96 (apple) Ekaterina Efstathiadi/Shutterstock; p98 (cake) qwe-store/Shutterstock; p100 (key) M_Videoues/Shutterstock; p102 (brick) Golden Sikorka/Shutterstock; p102 (feather) VectorXpert/Shutterstock; p105 (bath) Rvector/Shutterstock; p105 (bucket) yuwnis07/Shutterstock; p106 (digital clock) Alecsandr77/Shutterstock; p106 (watch) StockSmartStart/Shutterstock; p108 (shopping bag) Vector_Bird/Shutterstock; p108 (purse) ankomando/Shutterstock; p110, 111 (window) pikepicture/Shutterstock; p110, 111 (road sign) NEGOVURA31/Shutterstock; p111 (pizza) Lily Sab/Shutterstock; p111 (plate) baldezh/Shutterstock; p111 (wheel) WinWin artlab/Shutterstock; p112 (tissue box) DesignConcept/Shutterstock; p112 (can) lukpedclub/Shutterstock; p112 (carton) Valentina Vectors/Shutterstock; p112 (ice cream) Alena Divina/Shutterstock; p112 (pyramid) Giuseppe_R/Shutterstock

Contents

How to use 5
Words lists 7
100 square 8

English Made Simple 9

Sentences
Capital letters and full stops 10
Question marks 12
Exclamation marks 14
Capital letters for names and 'I' 16
Spaces between words 18
Sequencing words 20
Using 'and' 22
Sequencing sentences 24

Vocabulary
Adding 's' or 'es' 26
Adding 'ing' 28
Adding 'ed' 30
Adding 'er' and 'est' 32
Adding the prefix 'un' 34

Spelling
Days of the week 36
'nk' words 38
The 'ch' sound 40
Words ending in the 'v' sound 42
'ph' words 44
Words ending in 'y' 46
Contractions 48
Compound words 50
Tricky words (1) 52
Tricky words (2) 54

Reading
Titles, characters and events 56
Inference 58
Predict what happens next 60

Maths Made Simple 63

Number and place value
Counting forwards and backwards 64
Ordinal numbers 66
Counting in multiples of 2 68
Counting in multiples of 5 70
Counting in multiples of 10 72
One more, one less 74
Read and write numbers to 20 76

Addition and subtraction
Number bonds to 10 78
Adding numbers to 10.................................80
Subtracting numbers to 10 82
Number bonds to 20 84
Adding numbers to 20.................................86
Subtracting numbers to 20 88
Problem solving 90

Multiplication and division
Grouping 92
Sharing...94

Fractions
Halves 96
Quarters 98

Measurement
Length..100
Mass and weight 102
Capacity and volume 104
Time 106
Money 108

Geometry: Properties of shape
2D shapes..110
3D shapes..112

Geometry: Position and direction
Turns 114

English glossary 116
Maths glossary 117
English answers 119
Maths answers 122
Progress trackers 125

4

How to use

This book has been written to help children reinforce the English and maths skills they have learned in school. Each subject is divided into sections covering a range of topics from the National Curriculum. Use the book little and often to practise skills and increase confidence. You can choose to work through the English and maths sections in order or focus on specific topics.
At the back of the book is a **Progress tracker** to enable you to record what has been practised and achieved.

English

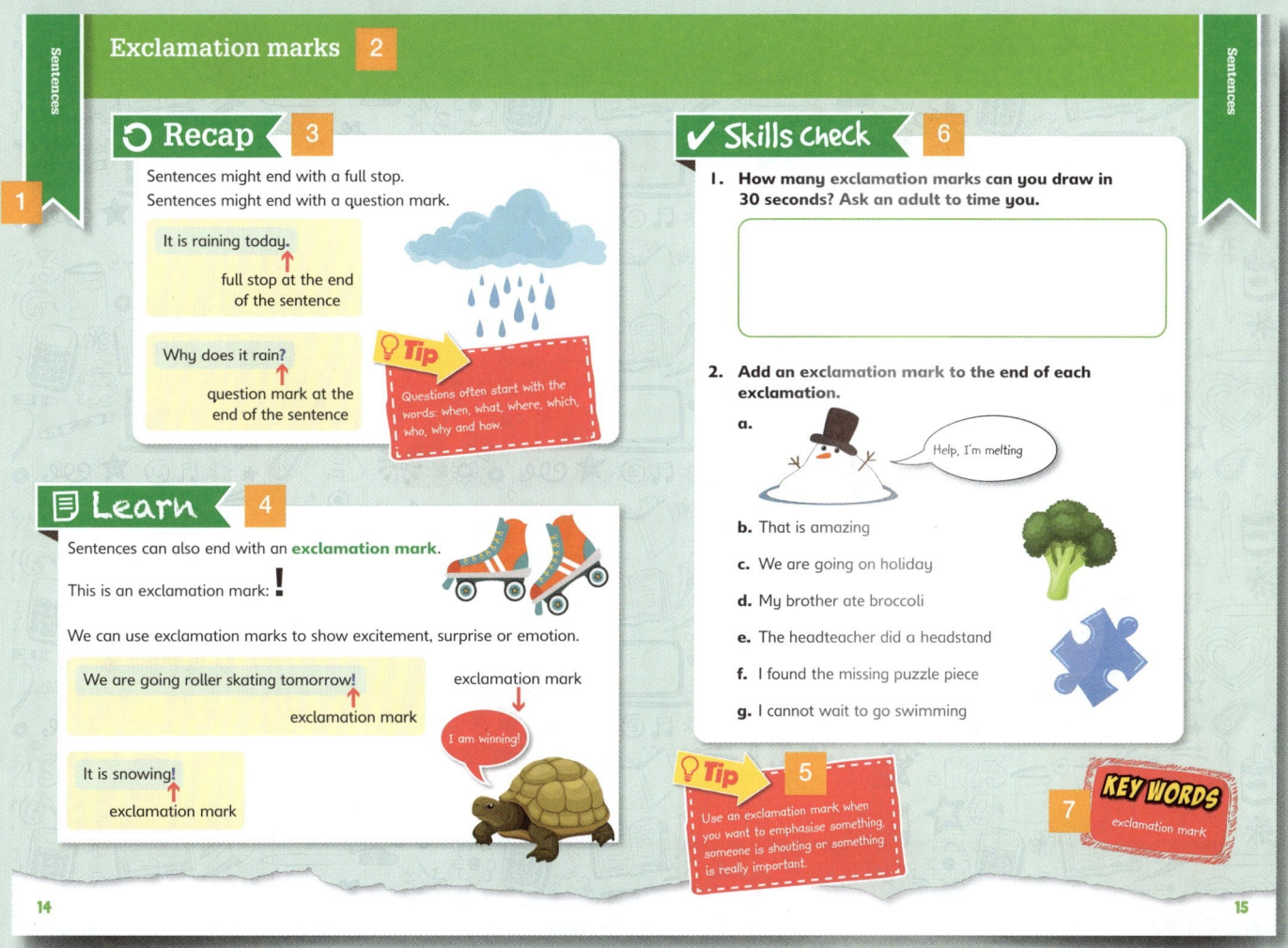

1. Chapter title
2. Topic title
3. Each page starts with a **recap** summarising what the children should know.
4. In the **learn** section there are clear explanations and examples, using illustrations and diagrams, where relevant.
5. **Tips** provide short and simple advice to aid understanding.
6. The **skills check** sections enable children to practise what they have learned with answers at the back of the book.
7. **Key words** that children need to know are displayed. Definitions for these words can be found in the **Glossary**.

Maths

The maths section has many of the same features as the English section and also some additional ones. Keep some blank or squared paper handy for notes and calculations!

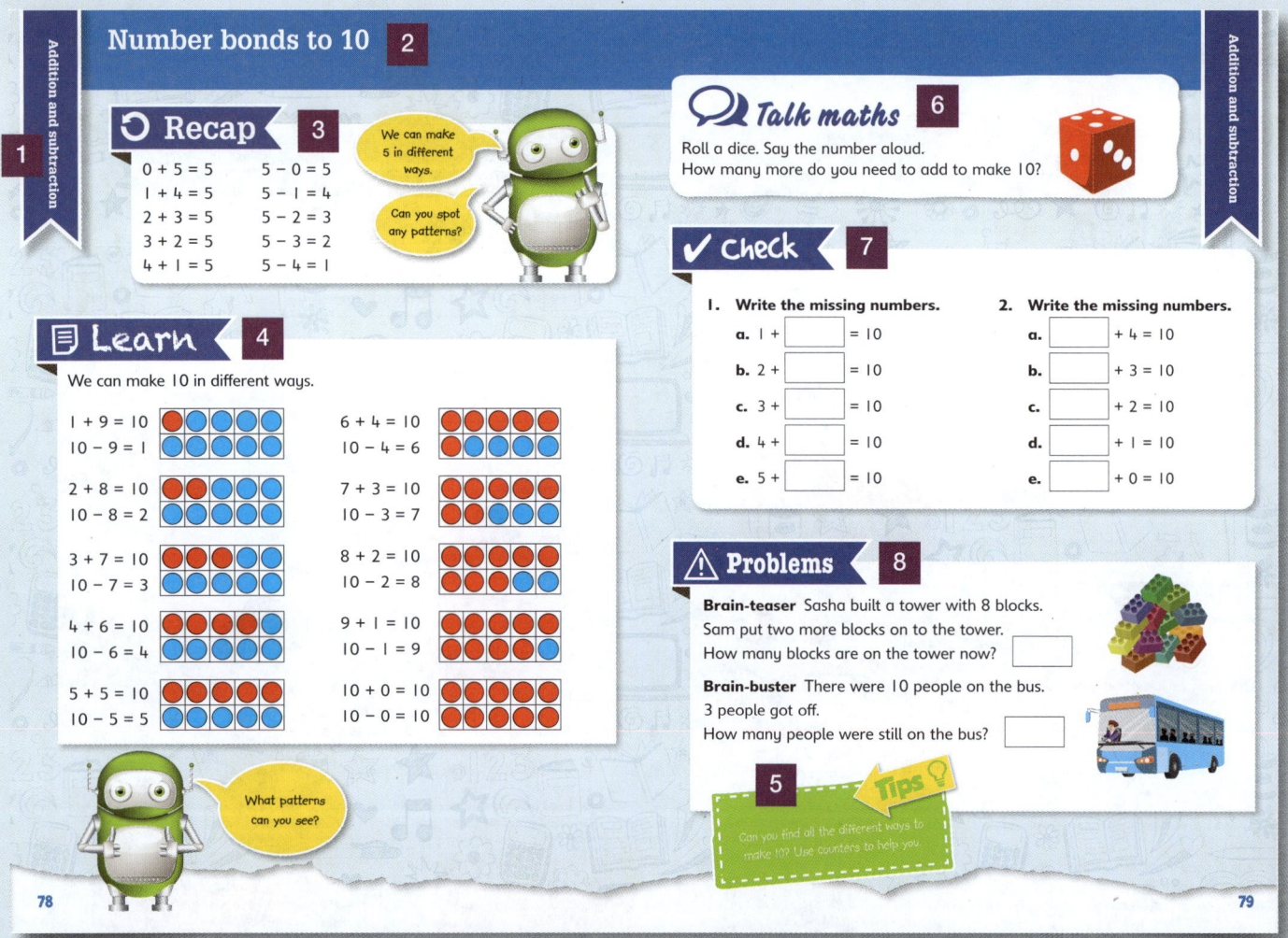

1. Chapter title
2. Topic title
3. Each page starts a **recap** of basic facts of the mathematical area in focus.
4. In the **learn** section there are clear explanations and examples, using illustrations and diagrams, where relevant.
5. **Tips** provide short and simple advice to aid understanding.
6. **Talk maths** are focused activities that encourage verbal practice.
7. **Check** a focused range of questions, with answers at the end of the book.
8. **Problems** word problems requiring mathematics to be used in context.

Tips for using this book at home
Using this book, alongside the maths and English being done at school, can boost children's mastery of the concepts. Be sure not to get ahead of schoolwork or to confuse your child.

Keep sessions to an absolute maximum of 30 minutes. Even if children want to keep going, short amounts of focused study on a regular basis will help to sustain learning and enthusiasm in the long run.

Word lists

These are the words you need to learn to spell.

Year 1

the	his	go	ask
a	has	so	friend
do	I	by	school
to	you	my	put
today	your	here	push
of	they	there	pull
said	be	where	full
says	he	love	house
are	me	come	our
were	she	some	
was	we	one	
is	no	once	

100 square

1	2	3	4	5	6	7	8	9	10
11	12	13	14	15	16	17	18	19	20
21	22	23	24	25	26	27	28	29	30
31	32	33	34	35	36	37	38	39	40
41	42	43	44	45	46	47	48	49	50
51	52	53	54	55	56	57	58	59	60
61	62	63	64	65	66	67	68	69	70
71	72	73	74	75	76	77	78	79	80
81	82	83	84	85	86	87	88	89	90
91	92	93	94	95	96	97	98	99	100

English Made Simple
Ages 5–6

Capital letters and full stops

 Recap

A **sentence** is a group of words that make sense.

What is a sentence?

Learn

Sentences start with a **capital letter** and end with a **full stop**.

Capital letters	A	B	C	D	E	F	G	H	I	J	K	L	M
Lower-case letters	a	b	c	d	e	f	g	h	i	j	k	l	m

Capital letters	N	O	P	Q	R	S	T	U	V	W	X	Y	Z
Lower-case letters	n	o	p	q	r	s	t	u	v	w	x	y	z

A full stop shows the reader the sentence has ended.

Tip: A full stop is a small dot.

capital letter at the beginning of the sentence

I like playing football**.**

full stop at the end of the sentence

capital letter at the beginning of the sentence

There is a rainbow in the sky**.**

full stop at the end of the sentence

KEY WORDS
sentence
capital letter
full stop

✓ Skills Check

1. **Circle the capital letters.**

   A   a   N   e   E   g   r   Q

   d   G   D   B   b   R   i   M

2. **Draw lines to match the capital letters to the lower-case letters.**

   | n | c | g | h | j | t |

   | J | T | N | C | H | G |

3. **Add a full stop to the end of each sentence.**

   a. The sun is shining

   b. I went to the park on Saturday

   c. I have a pet zebra and a whale lives in my bath

4. **Rewrite each sentence with a capital letter at the beginning and a full stop at the end.**

   a. we had pizza for lunch

   _____

   b. my coat is red

   _____

   c. the teacher told us a joke

   _____

Question marks

⟲ Recap

A sentence begins with a capital letter and ends with a full stop.

📄 Learn

When we ask something, it is called a question.

When we write a question down, we end with a **question mark**.

Here is a question mark: **?**

Why do people have different coloured eyes**?**

question mark at the end of the sentence

What colour is a giraffe's tongue**?**

question mark at the end of the sentence

A question mark shows the reader it's a question.

KEY WORDS

question mark

✓ Skills Check

How many question marks can you write in a minute?

1. **Draw a line of question marks.**

    ? ? ? _____

2. **Which of these are questions? Tick them.**

    I like chocolate ice cream. ☐

    Can an elephant run? ☐

    My favourite animal is an elephant. ☐

    How do you make ice cream? ☐

3. **Add a question mark to the end of each question.**

    a. Why is a cactus prickly

    b. Did dinosaurs have feathers

    c. How many stripes does a zebra have

4. **Write your own question. Remember to start with a capital letter and end with a question mark.**

    _____

    _____

💡 Tip

Here are some question words you can use to ask a question:

how where what when why which who

Can you make a question mark out of modelling dough?

Exclamation marks

↻ Recap

Sentences might end with a full stop.
Sentences might end with a question mark.

It is raining today.

↑ full stop at the end of the sentence

Why does it rain?

↑ question mark at the end of the sentence

💡 **Tip**

Questions often start with the words: when, what, where, which, who, why and how.

📄 Learn

Sentences can also end with an **exclamation mark**.

This is an exclamation mark: **!**

We can use exclamation marks to show excitement, surprise or emotion.

We are going roller skating tomorrow!

↑ exclamation mark

It is snowing!

↑ exclamation mark

exclamation mark ↓

I am winning!

✓ Skills Check

1. How many exclamation marks can you draw in 30 seconds? Ask an adult to time you.

2. Add an exclamation mark to the end of each exclamation.

   a.
   Help, I'm melting

   b. That is amazing

   c. We are going on holiday

   d. My brother ate broccoli

   e. The headteacher did a headstand

   f. I found the missing puzzle piece

   g. I cannot wait to go swimming

Tip
Use an exclamation mark when you want to emphasise something, someone is shouting or something is really important.

KEY WORDS

exclamation mark

Capital letters for names and 'I'

⟲ Recap

Here is the alphabet. The capital letters are in bold.

Capital letters	A	B	C	D	E	F	G	H	I	J	K	L	M
Lower-case letters	a	b	c	d	e	f	g	h	i	j	k	l	m

Capital letters	N	O	P	Q	R	S	T	U	V	W	X	Y	Z
Lower-case letters	n	o	p	q	r	s	t	u	v	w	x	y	z

We use capital letters at the beginning of a sentence.

📄 Learn

We use capital letters for people's names.

Alec   **M**aya   **B**en   **M**ax   **E**lla   **A**mira   **M**r **G**reen   **M**rs **J**ackson

We use a capital letter for the word **I**.

capital letter
↓
I like milkshake.   ✓
i like milkshake.   ✗

capital letter
↓
I am a superhero.   ✓
i am a superhero.   ✗

✓ Skills Check

1. Write the letters of the alphabet as capital letters.

Capital letters								

Capital letters								

Capital letters						

Capital letters				

2. Circle the words that should start with a capital.

a. ava went to the park.

b. mr green was a teacher.

c. Yesterday, max made a cake.

d. The lady who lives next door is called mrs jackson.

e. ben went to the park with jun.

f. alec and tom had ice cream after lunch.

3. Rewrite each sentence with the correct capital letters.

a. i like hopping like a grasshopper.

_____

_____

b. When it started raining, i splashed in the puddles.

_____

_____

Tip: We use capital letters for names, the word I and at the beginning of a sentence.

Sentences

Spaces between words

↻ Recap

A sentence is a group of words that makes sense.

📄 Learn

We put a small space between words when we are writing. This helps the reader to read the words more easily.

Remember the spaces are small!

Igrewasunflower. ✗
I grew a sunflower. ✓
I   grew   a   sunflower. ✗

✓ Skills Check

1. **Read these sentences aloud. Clap between each word.**

    a. Foxes hunt at night.

    b. Bees make honey.

    c. Plums are purple.

2. **Draw a line between the words in each sentence. The first one has been done for you.**

    a. I | had | an | apple | after | lunch.

    b. Giraffes have blue tongues.

    c. The snail slithered down the path.

3. **Copy the sentences. Remember to put a space between the words.**

a. Owls have big eyes.

_____

_____

b. There was a frog in the pond.

_____

_____

c. Some dinosaurs flew.

_____

_____

4. **Ava wrote a sentence. She forgot to put spaces between the words.**

Ilikepizza.

a. Draw lines between the words to show where the spaces should be.

b. Copy the sentence, remembering the spaces between the words.

_____

Tip

Put a lollipop stick, a piece of dried pasta, a pom pom or your finger after each word to practise leaving spaces.

When you are typing a sentence, you use the spacebar between words.

Sequencing words

🔄 Recap

A sentence is a group of words that makes sense.

When we write, we put a small space between each word.

small gaps between words

Sloths move slowly.

sentence

Sentences begin with a capital letter and end with a full stop.

📄 Learn

Sentences are made up from a group of words. The words must be in the right order to make sense.

train. on the I went ✗
I went on the train. ✓

💡 **Tip**

Say the sentence aloud to check it makes sense.

✓ Skills check

1. **Rearrange the words to make a sentence.**

    a. sweets.   I   like

    _____

    b. roared   lion   The   loudly.

    _____

    c. space.   rocket   into   went   The

    _____

    _____

    d. mud.   hippo   in   rolled   The

    _____

    _____

2. **Are the words in the correct order? Put a tick for yes and a cross for no.**

    a. sky. stars the shining are in The   ☐

    b. Ben rode his bike at the park.   ☐

    c. Cheetahs have spots.   ☐

    d. legs. have Snakes do not   ☐

Using 'and'

↻ Recap

Joining words join two words or two groups of words together. We can use the word **and** to join words together.

fish **and** chips
salt **and** pepper
girls **and** boys
the moon **and** stars

Think of a sentence using the word and.

📄 Learn

A sentence is a group of words that make sense.
We can use the word **and** to join sentences.

Zebras have stripes. Leopards have spots.

Zebras have stripes **and** leopards have spots.

I washed my face. I cleaned my teeth.

I washed my face **and** I cleaned my teeth.

Ava jumped in the puddles. Max danced in the rain.

Ava jumped in the puddles **and** Max danced in the rain.

✓ Skills Check

1. **Write the word 'and' to join the sentences together.**

    a. My bike is blue _____ Tom's bike is red.

    b. Pia is playing with the bricks _____ Jun is reading a book.

    c. Sam had an apple _____ Sasha had a banana.

2. **Join the sentences together using 'and'.**

    a. Cheetah are fast. Sloths are slow.

    _____
    _____

    b. Dogs bark. Cats purr.

    _____

    c. Elephants live on land. Whales live in the sea

    _____
    _____

💡 Tip
Remember to remove the full stop and change the capital letter of the second sentence (unless it's a name or I).

KEY WORDS
joining words

Sequencing sentences

↻ Recap

We order words to make sure a sentence makes sense.

The farmer grows corn.

📄 Learn

You can put sentences in order to make a story.
A story has a beginning, a middle and an end.

beginning — The dinosaur built a nest.

middle — The dinosaur laid an egg in the nest.

end — The egg hatched.

✓ Skills check

> Read the sentences again to check the order makes sense.

1. **Number the sentences to put them in the correct order. The first one has been done for you.**

   a. The dinosaur built a nest. ___1___

   The egg hatched. ___3___

   The dinosaur laid an egg in the nest. ___2___

   b. Then she baked fresh bread. _____

   The baker sold the fresh bread. _____

   The baker got up early to bake. _____

   c. It had snowed. _____

   The boy put a hat and a scarf on his snowman. _____

   The boy built a snowman. _____

   d. The gardener planted some seeds. _____

   The carrots grew. _____

   The gardener ate the carrots. _____

   e. She kicked the ball. _____

   She scored a goal. _____

   Ella put on her football boots. _____

🔄 Recap

What sound does the letter **s** make?

Say some words that begin with the letter **s**.

📄 Learn

A **suffix** is a letter or group of letters you add to the end of a word to change its meaning.

We add **s** or **es** to words to make them **plural**.

Singular means there is only one.   Plural means there is more than one.

cat   cats

For most words, we add **s**.

cat + **s** = cat**s**   door + **s** = door**s**

toy + **s** = toy**s**   car + **s** = car**s**

frog + **s** = frog**s**

If a word ends in **x**, **z**, **sh**, **ch** or **ss**, we add **es**.

fox + **es** = fox**es**   watch + **es** = watch**es**

buzz + **es** = buzz**es**   dress + **es** = dress**es**

dish + **es** = dish**es**

KEY WORDS

suffix
plural
singular

✓ Skills Check

1. **Add 's' to make these words plural.**

    a. hat_____

    b. tractor_____

    c. tree_____

    d. sock_____

    e. cake_____

    f. star_____

2. **Add 'es' to make these words plural.**

    a. box_____

    b. wish_____

    c. watch_____

    d. brush_____

    e. dress_____

Tip: Look at the end of the word to decide if it needs an s or es to make it plural.

🔄 Recap

A suffix is a letter or group of letters you add to the end of a word to change the meaning.

owl + **s** = owls

dish + **es** = dishes

📄 Learn

We can add the suffix **ing** to words.

jump + **ing** = jumping

I jump in puddles.
I am jumping in puddles.

cook + **ing** = cooking

I cook lunch.
I am cooking lunch.

draw + **ing** = drawing

I can draw a cat.
I am drawing a cat.

play + **ing** = playing

I play tennis.
I am playing tennis.

✓ Skills Check

1. **Add the suffix 'ing' to these words.**

   a. look + ing = _____

   b. paint + ing = _____

   c. help + ing = _____

   d. sing + ing = _____

   e. grow + ing = _____

2. **Choose the correct word to complete each sentence.**

   a. `look` `looking`

   I _____ out of the window.

   I am _____ out of the window.

   b. `paint` `painting`

   He is _____ the door.

   He can _____ doors.

   c. `help` `helping`

   I am _____ my mum to wash the car.

   I _____ Gran cook the dinner.

   d. `sing` `singing`

   I can _____ .

   I like _____ .

   e. `grow` `growing`

   Flowers _____ in the garden.

   We are _____ sunflowers.

 **Tip**

Say the sentence aloud to check it makes sense.

🔄 Recap

A suffix is a letter or group of letters you add to the end of a word to change the meaning. **s**, **es** and **ing** are suffixes.

What is a suffix?

📄 Learn

We can add the suffix **ed** to some words to change it into the **past tense**.

💡 Tip
Something in the past tense has already happened.

walk + **ed** = walked

I walk up the hill.
I walked up the hill.

play + **ed** = played

I play tennis.
I played tennis.

wash + **ed**

I wash the dishes.
I washed the dishes.

✓ Skills Check

1. **Colour in the suffix 'ed' in these words.**

   a. w a l k e d   e. j u m p e d

   b. p l a y e d   f. c o o k e d

   c. h e l p e d   g. w a s h e d

   d. l o o k e d   h. p u l l e d

2. **Add the suffix 'ed' to these words.**

   a. smash_____   f. thump_____

   b. bang_____   g. crash_____

   c. push_____   h. wish_____

   d. pick_____   i. buzz_____

   e. clean_____

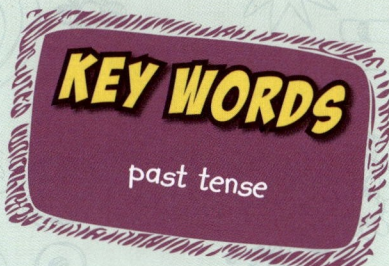

KEY WORDS

past tense

Suffixes are added to the end of a word.

🔄 Recap

Suffixes are a letter or group of letters. Suffixes are added to the end of a word to change its meaning.

What is a suffix?

📋 Learn

We can add the suffixes **er** and **est** to some words.

tall

tall + **er** = taller

tall + **est** = tallest

The red flower is the tallest.
The yellow flower is the shortest.

The yellow flower is tall.
The orange flower is taller.
The red flower is the tallest.

small

small + **er** = smaller

small + **est** = smallest

The kitten is small.
The bird is smaller.
The mouse is the smallest.

kitten   bird   mouse

✓ Skills Check

1. **Colour in the suffix for each word.**

   a. s m a l l e r   s m a l l e s t

   b. l o n g e r   l o n g e s t

   c. o l d e r   o l d e s t

   d. n e w e r   n e w e s t

2. **Add the suffix to these words.**

   a. short + er = _____

   short + est = _____

   b. high + er = _____

   high + est = _____

3. **Circle the tallest building.**

> **Tip**
> We use er to compare two things. We use est to compare three or more things.

Adding the prefix 'un'

Recap

When you add a **prefix** to a word its meaning changes.

What is a prefix?

Learn

A prefix is a letter or group of letters you add to the start of a word to change its meaning.

un is a prefix.

When **un** is added to a word the new word means the opposite.

| happy | unhappy | unhappy means **not** happy |

| tie | untie | untie means **not** tied |

| fold | unfold | unfold means **not** folded |

| kind | unkind | unkind means **not** kind |

✓ Skills Check

1. Add the prefix 'un' to these words.

   a. _____ happy
   b. _____ tie
   c. _____ do
   d. _____ block

   e. _____ fold
   f. _____ cover
   g. _____ zip
   h. _____ kind

2. Can you find the words with the prefix 'un' in the wordsearch?

   uncover   untie   unfold   undo   unzip

u	n	c	o	v	e	r	n	v	w
h	l	p	w	q	u	s	a	c	v
c	b	n	a	y	t	u	t	u	l
u	n	t	i	e	u	n	l	n	b
t	i	o	q	j	o	f	l	d	p
d	c	b	n	n	b	o	l	o	b
u	n	z	i	p	n	l	t	r	w
q	w	r	t	y	u	d	g	h	j

The prefix un means not.

Tip

Prefixes are added to the start of a word. Suffixes are added to the end of a word.

KEY WORDS

prefix

Days of the week

↻ Recap

There are seven days in a week.
The days of the week always start with a **capital letter**.

📄 Learn

Here are the days of the week:
Monday, Tuesday, Wednesday, Thursday, Friday, Saturday, Sunday

Practise saying them aloud in order.

✓ Skills Check

1. **Can you find the days of the week in the wordsearch?**

   Monday   Tuesday   Wednesday   Thursday
   Friday   Saturday   Sunday

M	M	O	N	D	A	Y	Y	T	C
D	A	R	T	M	O	N	Y	U	S
T	H	U	R	S	D	A	Y	E	A
T	L	V	K	B	S	N	N	S	T
M	S	U	N	D	A	Y	Y	D	U
W	N	O	M	R	Y	V	B	A	R
W	E	D	N	E	S	D	A	Y	D
E	R	T	Y	G	H	C	A	T	A
D	R	F	R	I	D	A	Y	P	Y

Do you know what day it is today? What day will it be tomorrow?

36

2. Copy the days of the week.

Monday _____

Tuesday _____

Wednesday _____

Thursday _____

Friday _____

Saturday _____

Sunday _____

KEY WORDS

capital letters

3. Fill in the missing letters.

M	o	n		a	y		
T		e	s	d	a	y	
W	e	d	n	e	s			
		u	r	s	d	a	y
			d	a	y		
S	a	t			d	a	y
			d	a	y		

4. Clap the days of the week.

Mon/day 👏👏

Tues/day 👏👏

Wednes/day 👏👏

Thurs/day 👏👏

Fri/day 👏👏👏

Sat/ur/day 👏👏👏

Sun/day 👏👏

Tip

Breaking the words up into chunks can help you remember how to spell them.

↻ Recap

We can sound out words to help us read them.

p i nk	th i nk	b a nk
. . _	_ . _	. . _

📄 Learn

nk

Here are some words with the **nk** sound. Have a go at saying the words aloud.

This sound is usually found at the end of words.

pink	thank
think	sank
sink	drank
drink	tank
link	bank
wink	honk

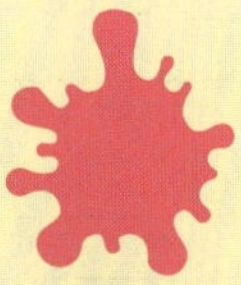

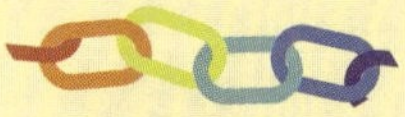

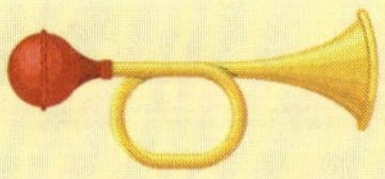

✓ Skills check

1. **Add the missing letters to complete these words.**

    a.  thi_____

    b.  li_____

    c.  ta_____

    d.  ba_____

2. **Add the missing letters to complete these words.**

    a. The boat s_____ to the bottom of the sea.

     b. My shoes are p_____.

    c. Can I have a d_____ please?

     d. Wash your hands at the s_____.

Recap

We can **segment** words to spell them.
We can **blend** words to read them.

l u n ch
lunch

Learn

Let's look at words with the ch sound. This sound can be spelled ch or tch.

ch

r	i	ch		rich
m	u	ch		much
ch	i	ck		chick
ch	i	n		chin

Have a go at segmenting and blending the words.

tch

h	u	tch		hutch
c	a	tch		catch
d	i	tch		ditch
p	a	tch		patch

Read the sentences. Can you find the words with the **ch** sound?

Can the dog fetch the ball?

We are cooking in the kitchen.

It was such a wonderful day.

Which way is it to the park, please?

✓ Skills Check

1. **Colour in the ch sound in each word.**

   a. | h | a | t | c | h |

   b. | c | h | i | p |

   c. | l | a | t | c | h |

   d. | c | h | e | s | t |

   e. | s | n | a | t | c | h |

   f. | i | t | c | h |

2. **Circle the correct spelling for each word.**

   a. match   mach

   b. bench   bentch

   c. cruntch   crunch

   d. pitch   pich

   e. churtch   church

   f. pinch   pintch

Tip

When you split the words into their sounds it is called segmenting. When you blend the sounds together to read a word, it is called blending.

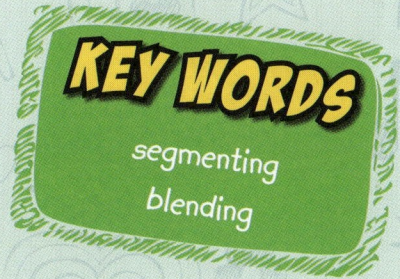

KEY WORDS

segmenting
blending

↻ Recap

van   vet   vest

📋 Learn

Words that end in the **v** sound are spelled with a **v** and an **e**.

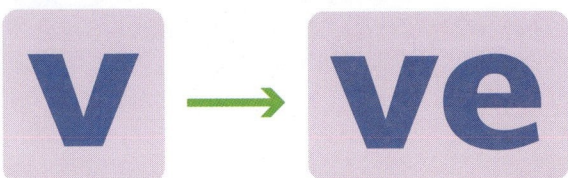

| thieve | massive | carve | solve | leave |
| halve | serve | shove | prove | halve |

✓ Skills Check

1. **Can you find the words that end with the v sound?**

h	d	s	t	b	c	v	w
a	n	l	p	j	h	t	r
v	r	e	a	b	o	v	e
e	t	e	k	w	r	t	n
n	w	v	n	v	v	c	r
l	t	e	m	g	i	v	e
m	o	v	e	k	l	n	m
p	d	t	v	a	l	v	e

have
sleeve
give

move
above
valve

2. **Circle the words that are spelled correctly.**

shelve   shelv   curve   starve   starv

activ   active   massive   nerv   weave

weav   curv   nerve   move   massiv

3. **Add the missing letters to complete these words.**

a. glo_____

b. oli_____

c. lo_____

d. do_____

↻ Recap

fish 　　feet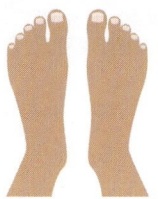

fairy 　　frog

Read these words beginning with the f sound.

📄 Learn

ph makes the same sound as **f**

dolphin 　　alphabet

sphere 　　phone

elephant 　　trophy

✓ Skills Check

1. **Colour in the ph sounds in the words.**

   a. | d | o | l | p | h | i | n |

   b. | s | p | h | e | r | e |

   c. | e | l | e | p | h | a | n | t |

   d. | p | h | o | n | e |

   e. | t | r | o | p | h | y |

   f. | a | l | p | h | a | b | e | t |

2. **Sort the words into the correct box.**

   dolphin   fairy   feet   trophy   phone
   frog   fun   first alphabet sphere

   f

   _____
   _____
   _____
   _____
   _____

   ph

   _____
   _____
   _____
   _____
   _____

Recap

yak   yellow   yoghurt

Read the words beginning with y.

Learn

y at the end of a word makes an **ee** sound.

happy

cherry

bunny

lady

puppy

party

berry

family

✓ Skills Check

1. **Add the y to the end of each word.**

   a. happ____

   b. cherr____

   c. bunn____

   d. lad____

   e. pupp____

   f. part____

   g. berr____

   h. famil____

2. **Circle the words that are spelled correctly.**

   happee   lady   bunny   puppe   parte

   berry   happy   famile   family   bunee

3. **Find the words that end in y.**

   happy
   lady
   cherry

h	a	p	p	y	n	y	c
g	f	g	h	w	r	t	h
w	r	t	h	l	n	m	e
w	c	v	k	a	e	r	r
r	s	b	v	d	l	l	r
b	e	r	r	y	p	p	y
t	b	p	a	r	t	y	g
f	a	m	i	l	y	n	v

   family
   party
   berry

🔄 Recap

We use a capital letter for the word **I**.

I like apples. How will **I** get to the park?
Ben and **I** played football.

📄 Learn

This is called an **apostrophe**.

It shows a letter or letters are missing.

Sometimes we use apostrophes to shorten words. These words are called **contractions**.

I'll = I will

I'm = I am

I've = I have

I'd = I would

we'll = we will

we've = we have

What letters are missing in these contractions?

✓ Skills Check

1. **Draw a line to match each word to the correct contraction.**

I would	we'll
I am	I'm
I will	I've
we will	we've
we have	I'll
I have	I'd

2. **Circle the contractions in these sentences.**

   a. I'm going to sing a song on the stage.

   b. I've got a brother.

   c. We'll have to walk to school.

   d. We've got half an hour until the next bus.

   e. I'll wait for you.

   f. I'd like a drink.

KEY WORDS
apostrophe
contraction

↻ Recap

Read these words.

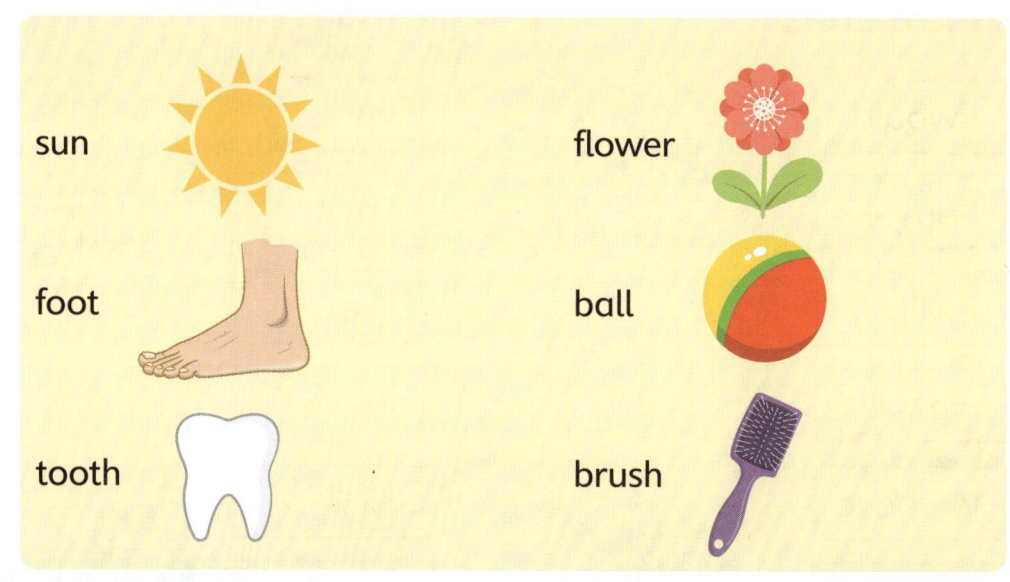

sun

flower

foot

ball

tooth

brush

🗐 Learn

You can add two words together to make a new word.

tooth + brush = toothbrush

snow + ball = snowball

sun + flower = sunflower

foot + ball = football

pop + corn = popcorn

eye + brow = eyebrow

💡 Tip

Compound words are made from two other words joined together.

✓ Skills Check

1. **Draw a line to join two words together and make a new word.**

tooth		corn
snow		berry
sun		man
pop		shine
straw		paste

2. **Add another word to create a compound word. Use the words in the box to match each picture.**

   | flake | drop | berry | mat |

   a. rain_____

   b. snow_____

   c. blue_____

   d. door_____

How many compound words can you think of?

Tricky words (1)

Recap

What sounds do these letters make?

p**e**t
t**e**n
r**e**d
c**o**t
b**o**x
s**o**b
yak
yoghurt
yell

Learn

Some spellings don't follow the rules!

Have a go at reading and spelling these words:

In these words the y makes an igh sound.

by   my

In these words the e makes an ee sound.

be   he   she   we

In these words the o makes an oa sound.

go   so   no

Spelling

✓ Skills check

1. Write a line of each word.

    a. be _____

    b. he _____

    c. she _____

    d. we _____

    e. go _____

    f. so _____

    g. no _____

    h. by _____

    i. my _____

2. Choose a word from Question 1. How many times can you write it in 30 seconds? Ask an adult to time you.

> **Tip**
>
> Practise writing these words in fun ways. Can you write them in chalk? Can you make them out of modelling dough? Can you create them using letter magnets? How big can you write each word? How small can you write each word?

Spelling

Tricky words (2)

↻ Recap

Learn to read and spell these words off by heart.

Some words don't follow spelling rules.

📄 Learn

We use these words a lot when talking or reading and writing.

Here are some more words that don't follow the rules!

the	today	said
are	was	you
there	where	one
school	house	some

✓ Skills Check

1. Find the words in the wordsearch.

the									there
today	s	s	o	m	e	t	h	w	where
said	c	a	t	o	d	a	y	a	one
are	h	r	k	l	h	g	d	s	school
was	o	e	s	a	i	d	w	r	house
you	o	h	o	u	s	e	y	b	some
	l	n	o	n	e	h	o	t	
	m	t	h	e	r	e	u	h	
	w	h	e	r	e	p	n	e	

54

2. Copy these words.

a. the _____

b. today _____

c. said _____

d. are _____

e. was _____

f. you _____

g. there _____

h. where _____

i. one _____

j. school _____

k. house _____

3. Fill in the missing words. Use the words in the box.

| said | house | One | was |

Max lives in the _____ with a red door.

_____ day a robot came to school.

There _____ a robot on the playground.

"I can see a robot," _____ Max.

Tip

Try making up a saying using the letters of a word to help you remember how to spell the word. For example:

was: **w**hales **a**re **s**uper!

said: **s**nakes **a**re **i**mpressive **d**ancers!

Titles, characters and events

↻ Recap

A **sentence** is a group of words that makes sense.

📄 Learn

Titles are headings that tell you what the piece of writing is about or what a story is called.

Characters are the people or animals in a story.

Events are things that happen in a story or report.

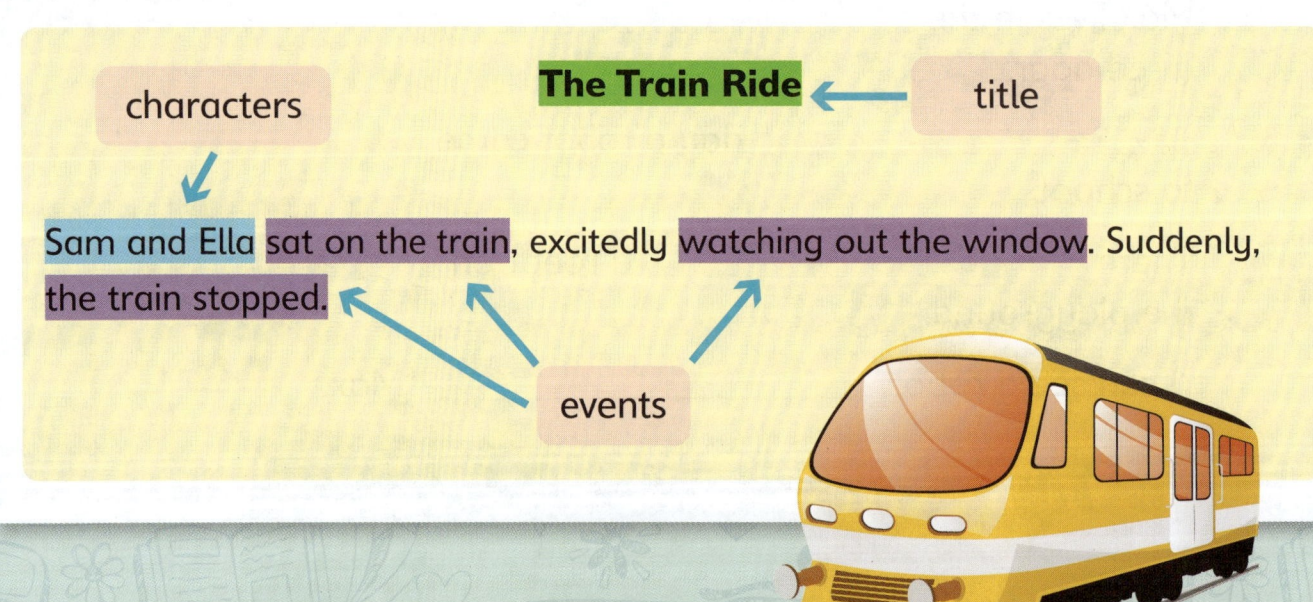

✓ Skills Check

1. **Read the text. Answer the questions.**

   A Visit to the Vets

   Pia and her nan were going to the vets.
   Pia's bunny needed a check-up.
   But the bunny did not want to get in the box.

   a. What is the title?

   _____

   b. Who are the two characters?

   _____
   _____

   c. What are the three events?

   _____
   _____
   _____
   _____
   _____
   _____

2. **Read the text. Circle the title. Underline the characters. Talk about the events.**

   At the Park

   Dad took Max to the park. Max played in the sandpit with his truck.

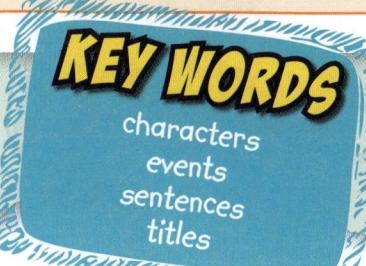

KEY WORDS
characters
events
sentences
titles

Inference

↻ Recap

We can find facts in a text.

> The owl had white feathers and big yellow eyes.

What colour are the owl's feathers? What size are the owl's eyes?

📄 Learn

We read text for meaning. Sometimes the answers are in the text:

> The cupcake lay on the floor at the boy's feet. Bits of mud and grass were stuck to the pink icing on the cupcake. Tears dripped down the boy's face.

What was on the floor? — *A cupcake!*

What was stuck to the cupcake? — *Mud and grass.*

Sometimes we can work out what has happened even when it doesn't say directly in the text, by looking for clues.

What do you think had happened to the cupcake? — *I think the boy dropped his cupcake because it is on the floor by his feet.*

How is the boy feeling? — *I think the boy is sad because he is crying.*

✓ Skills Check

Read the text and answer the questions.
Tick the correct answers.

Sam and his grandad were going fishing. Excitedly, Sam helped his grandad put the fishing rods in the car. "Can you get my fishing chair out of the shed?" asked Grandad. Sam looked for the chair in the shed. He saw a spider so he screamed and ran back to the car.

1. Where are Sam and his grandad going?

   swimming ☐

   fishing ☐

   running ☐

2. Where was Grandad's chair?

   in the kitchen ☐

   in the car ☐

   in the shed ☐

3. How did Sam feel about spiders?

   scared ☐

   excited ☐

   happy ☐

To infer something means to work it out from the information given.

Tip

When we look for clues in the text to answer questions, it is called **inference**. We infer something from the text.

KEY WORDS

inference

Predict what happens next

↻ Recap

What does predict mean?

Predict means to guess what happens next.

📄 Learn

We can use clues in a story to predict what might happen next.

Pia left her bike on the floor while she got an ice cream. The ice cream was big and it was a little bit wobbly. Pia saw her friend, Fred, and waved to him but she was not looking where she was going.

What do you think will happen next?

Here are the clues:

Pia left her bike on the floor ⟶ This tells us there is something to trip over.

The ice cream was big and wobbly. ⟶ This tells us the ice cream can fall easily.

Pia was not looking where she was going. ⟶ This tells us Pia will not see the bike.

I predict Pia is going to trip over the bike and drop her ice cream.

Can you explain your answer?

I predict this because Pia left her bike on the floor and she is not looking where she is going.

✓ Skills Check

Read the text.

The New Slide

Ben went to the park. Ben liked the swings best. He did not like being high up or going fast. Today, there was a new slide at the park. The new slide was very high and fast. Ben went up all the steps to the new slide very slowly. When he got to the top, he looked down. It was a very long way – Ben was not very happy.

Can you predict what happens next?

1. What do you think happened next? Tick your answer.

   Ben does not want to go down the slide. ☐

   Ben likes the slide because it is fast. ☐

   Ben likes the slide because it is high. ☐

2. Explain why you chose the answer you did.

   _____
   _____
   _____

> 💡 **Tip**
> Look for all the clues as you read the text. Highlight the clues to help you.

KEY WORDS

predict

Maths Made Simple
Ages 5–6

Counting forwards and backwards

↻ Recap

We can count forwards.
0, 1, 2, 3, 4, 5, 6, 7, 8, 9, 10

We can count backwards.
10, 9, 8, 7, 6, 5, 4, 3, 2, 1, 0

We can count objects.

There are eight planets.

Tips: The order of the numbers always stays the same.

I can count forwards starting at 18.
18, 19, 20, 21, 22, ...

📄 Learn

When you count forwards on a number line, the numbers get bigger. You can start counting from any number.

0 1 2 3 4 5 6 7 8 9 10 11 12 13 14 15 16 17 ⑱ 19 20 21 22 23 24 25 26 27 28 29 30

When you count backwards on a number line, the numbers get smaller. You can start counting from any number.

0 1 2 3 4 5 6 7 8 9 10 11 12 13 14 15 16 17 ⑱ 19 20 21 22 23 24 25 26 27 28 29 30

You can count more than 10 objects.
How many stars are there?

★★★★★★★★★★★★★★★

I can count backwards starting at 18.
18, 17, 16, 15, 14, ...

💬 Talk maths

Use the number line to count aloud.

Start at 15. Count forwards aloud.

Start at 15. Count backwards aloud.

✓ Check

1. **Count forwards.**

   a. Start at 10. Count forwards. What numbers are next?

   10, ☐, ☐, ☐, ☐, ☐

   b. Start at 22. Count forwards. What numbers are next?

   22, ☐, ☐, ☐, ☐, ☐

2. **Count backwards.**

   a. Start at 9. Count backwards. What numbers are next?

   9, ☐, ☐, ☐, ☐, ☐

   b. Start at 26. Count backwards. What numbers are next?

   26, ☐, ☐, ☐, ☐, ☐

3. **How many stars are there?**

   There are

   ★★★★★★★★★★★★★★★★★★★★★★★★ ☐ stars

⚠ Problems

Brain-teaser: Max is counting the steps as he goes up the stairs. What are the next numbers?

27, ☐, ☐, ☐, ☐, ☐

Can you count to 100?

Ordinal numbers

↻ Recap

There are four children running.

How many children are there?

🗐 Learn

Who is first?

Maya is in 1st place

We use **ordinal numbers** to say what order things are in.

1st – first	6th – sixth
2nd – second	7th – seventh
3rd – third	8th – eighth
4th – fourth	9th – ninth
5th – fifth	10th – tenth

🗨 Talk maths

Say the ordinal numbers aloud.

1st – first
2nd – second
3rd – third
4th – fourth
5th – fifth

6th – sixth
7th – seventh
8th – eighth
9th – ninth
10th – tenth

✓ Check

1. **Draw lines to match the ordinal numbers to the words.**

   | 1st | 2nd | 3rd | 4th | 5th | 6th | 7th | 8th | 9th | 10th |

   | tenth | second | seventh | first | fifth |
   | eighth | third | ninth | fourth | sixth |

2. **Rewrite the ordinal numbers in order.**

   5th   2nd   4th   1st   3rd

   ☐   ☐   ☐   ☐   ☐

⚠ Problems

Brain-teaser: Ava comes 2nd in a skipping race. Tom is next. What place is Tom in?

Tom is in ☐ place.

Counting in multiples of 2

↻ Recap

Here are 10 socks. We can count the socks one by one.

📄 Learn

Count in 2s.

0, 2, 4, 6, 8, 10

There are 10 socks.

0, 2, 4, 6, 8, 10, 12, 14, 16

There are 16 wheels.

Use a number line to count in 2s.

Follow the jumps with your finger.

💬 Talk maths

Count in 2s aloud.

0, 2, 4, 6, 8, 10, 12, 14, 16, 18, 20

What do you notice about these numbers?

✓ Check

1. **Count in 2s. Fill in the missing numbers.**

   a. 0, 2, 4, 6, ☐, 10

   b. 0, 2, ☐, ☐, 8, 10

   c. 0, ☐, 4, 6, 8, ☐

   d. 8, 10, 12, ☐, 16, 18, 20

   e. 8, 10, ☐, ☐, ☐, 18, 20

   f. ☐, 10, 12, 14, 16, ☐, 20

2. **Start at 6. Count forwards in steps of 2. Write down all the numbers below 20.**

   The first two steps have been done for you.

   6, 8, ☐, ☐, ☐, ☐, ☐, 20

⚠ Problems

Brain-teaser: There are 6 hens.
Each hen lays 2 eggs.
Count the eggs in 2s.
How many eggs are there?

Use a number line to help you practise counting in 2s.

 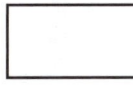    ☐

Counting in multiples of 5

↻ Recap

We can count to 5.
There are 5 fingers.

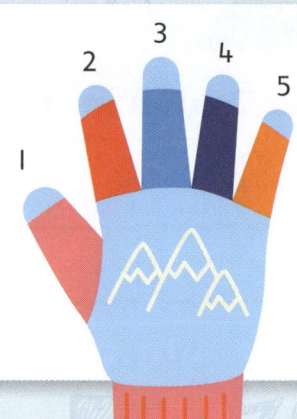

🗐 Learn

Count in 5s.

0, 5, 10, 15, 20, 25, 30

There are 30 fingers.

0, 5, 10, 15, 20, 25, 30, 35, 40

There are 40 petals.

Use a number line to count in 5s.

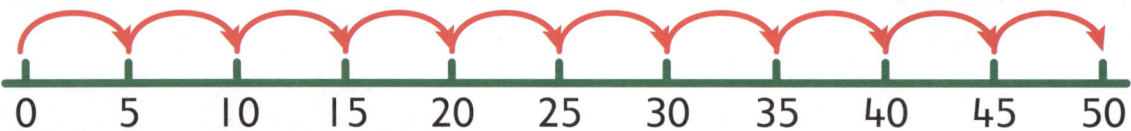

Follow the jumps with your finger.

💬 Talk maths

Do you notice anything about these numbers?

Count in 5s aloud.

0, 5, 10, 15, 20, 25, 30, 35, 40, 45, 50

✓ Check

1. **Count in 5s. Fill in the missing numbers.**

    a. 0, 5, 10, 15, 20, ☐, 30

    b. 0, 5, ☐, 15, 20, 25, 30

    c. 0, 5, 10, 15, ☐, 25, 30

    d. 20, 25, ☐, 35, 40, 45, 50

    e. 20, 25, 30, 35, 40, ☐, 50

    f. 20, ☐, 30, ☐, 40, ☐, 50

⚠ Problems

Brain-teaser: There are 4 bags. Each bag has 5 apples.

Count in 5s.

How many apples are there altogether? ☐

Brain-teaser: Delun has 6 coins.

Count in 5s.

How many pennies does Delun have? ☐ p

Tips: Use a number line to help you practise counting in 5s.

Counting in multiples of 10

↻ Recap

1	2	3	4	5	6	7	8	9	10
11	12	13	14	15	16	17	18	19	20
21	22	23	24	25	26	27	28	29	30
31	32	33	34	35	36	37	38	39	40
41	42	43	44	45	46	47	48	49	50
51	52	53	54	55	56	57	58	59	60
61	62	63	64	65	66	67	68	69	70
71	72	73	74	75	76	77	78	79	80
81	82	83	84	85	86	87	88	89	90
91	92	93	94	95	96	97	98	99	100

Count to 100.

📄 Learn

Count in 10s.

0, 10, 20, 30, 40, 50

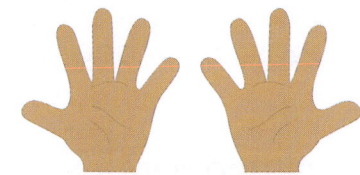

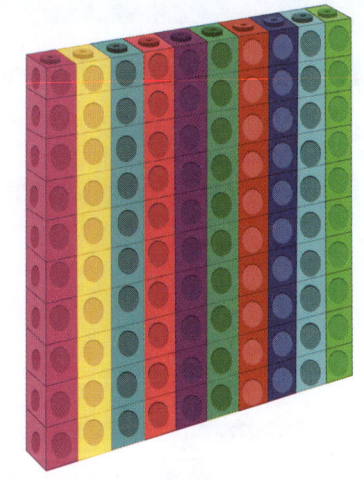

0, 10, 20, 30, 40, 50, 60, 70, 80, 90, 100

Use a number line to count in 10s.

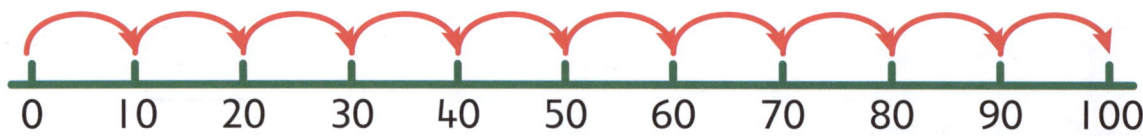

What do you notice about the tens?

💬 Talk maths

Count aloud in 10s.
Use the hundred square to help you.

1	2	3	4	5	6	7	8	9	10
11	12	13	14	15	16	17	18	19	20
21	22	23	24	25	26	27	28	29	30
31	32	33	34	35	36	37	38	39	40
41	42	43	44	45	46	47	48	49	50
51	52	53	54	55	56	57	58	59	60
61	62	63	64	65	66	67	68	69	70
71	72	73	74	75	76	77	78	79	80
81	82	83	84	85	86	87	88	89	90
91	92	93	94	95	96	97	98	99	100

✓ Check

What patterns can you see when counting in 10s?

1. Count in 10s. Fill in the missing numbers.

   a. 0, 10, ☐, 30, 40, 50

   b. 0, 10, 20, 30, ☐, 50

   c. 0, 10, 20, 30, 40, ☐

   d. 60, ☐, 80, 90, 100

   e. 60, 70, ☐, 90, 100

   f. 60, 70, 80, ☐, 100

   g. 0, ☐, 20, ☐, 40, ☐, 60, ☐, 80, ☐, 100

   h. 0, 10, ☐, 30, ☐, 50, ☐, 70, ☐, 90, ☐

⚠ Problems

Brain-teaser: Crayons are sold in boxes of 10.
Mr Field buys 4 boxes of crayons.
How many crayons does he have?

☐

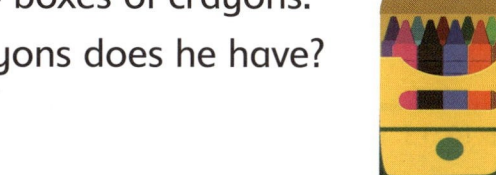

One more, one less

🔄 Recap

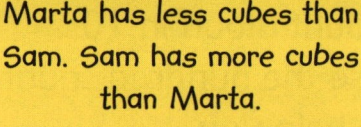

Marta has less cubes than Sam. Sam has more cubes than Marta.

Marta has 6 cubes.   Sam has 7 cubes.

📄 Learn

To find one less, take away one. To find one more, **add** one.

Find one less than 6.
6 – 1 = 5

Find one more than 6.
6 + 1 = 7

You can also use a number line to find one less and one more.

Find one less than 18.
18 – 1 = 17

0 1 2 3 4 5 6 7 8 9 10 11 12 13 14 15 16 17 18 19 20 21 22 23 24 25 26 27 28 29 30

Find one more than 18.
18 + 1 = 19

0 1 2 3 4 5 6 7 8 9 10 11 12 13 14 15 16 17 18 19 20 21 22 23 24 25 26 27 28 29 30

🗨 Talk maths

Roll a dice.

Find one more than the number on the dice.

Find one less than the number on the dice.

Try using two dice. Add the numbers together. Then find one more and one less than the total.

✓ Check

1. **Find one more and one less.**

   a. One more than 5 = ☐   one less than 5 = ☐

   b. One more than 15 = ☐   one less than 15 = ☐

   c. One more than 25 = ☐   one less than 25 = ☐

Do you notice a pattern?

⚠ Problems

Brain-teaser: The baker had 16 cakes. He sold one.

How many were left? ☐

Brain-buster: There were 29 children in the class.

One more child joined the class.

How many were in the class now? ☐

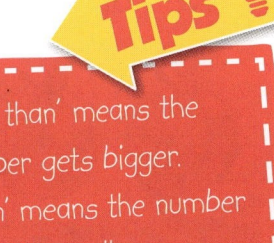

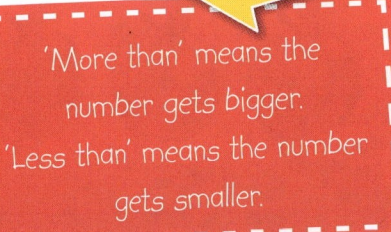

'More than' means the number gets bigger. 'Less than' means the number gets smaller.

Read and write numbers to 20

↻ Recap

We can show numbers in different ways.

📄 Learn

We can write numbers in numerals and in words.

1	one	2	two	3	three	4	four
5	five	6	six	7	seven	8	eight
9	nine	10	ten	11	eleven	12	twelve
13	thirteen	14	fourteen	15	fifteen	16	sixteen
17	seventeen	18	eighteen	19	nineteen	20	twenty

7 apples

seven apples

💬 Talk maths

Have a go at reading the numbers aloud:

one, two, three, four, five, six, seven, eight, nine, ten, eleven, twelve, thirteen, fourteen, fifteen, sixteen, seventeen, eighteen, nineteen, twenty.

✓ Check

1. Draw lines to match the numbers to the words.

4		thirteen
7		twenty
9		seven
13		four
17		nine
20		seventeen

⚠ Problems

Brain-teaser: Tamsin lives at number sixteen.
Which house is Tamsin's?
Draw a circle around the correct door.

Tips 💡

Look out for numbers in numerals and words around the house or at the shops. What numbers can you see? Have a go at reading them.

Recap

0 + 5 = 5   5 − 0 = 5
1 + 4 = 5   5 − 1 = 4
2 + 3 = 5   5 − 2 = 3
3 + 2 = 5   5 − 3 = 2
4 + 1 = 5   5 − 4 = 1

We can make 5 in different ways.

Can you spot any patterns?

Learn

We can make 10 in different ways.

1 + 9 = 10
10 − 9 = 1
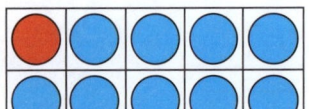

2 + 8 = 10
10 − 8 = 2
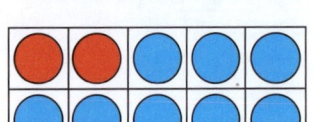

3 + 7 = 10
10 − 7 = 3
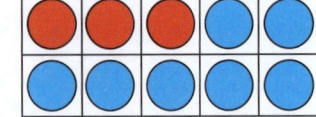

4 + 6 = 10
10 − 6 = 4
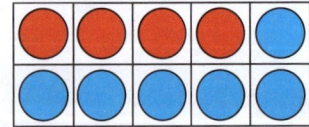

5 + 5 = 10
10 − 5 = 5
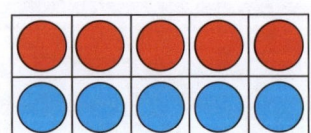

6 + 4 = 10
10 − 4 = 6
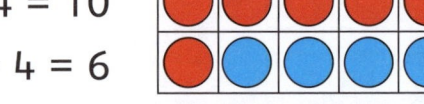

7 + 3 = 10
10 − 3 = 7
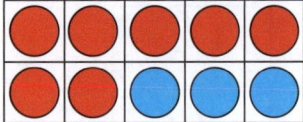

8 + 2 = 10
10 − 2 = 8
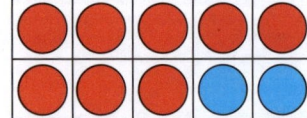

9 + 1 = 10
10 − 1 = 9
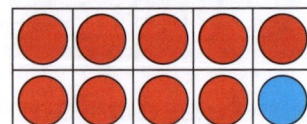

10 + 0 = 10
10 − 0 = 10
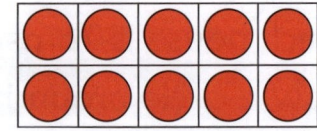

What patterns can you see?

💬 Talk maths

Roll a dice. Say the number aloud.
How many more do you need to add to make 10?

✓ Check

1. Write the missing numbers.

a. 1 + ☐ = 10

b. 2 + ☐ = 10

c. 3 + ☐ = 10

d. 4 + ☐ = 10

e. 5 + ☐ = 10

2. Write the missing numbers.

a. ☐ + 4 = 10

b. ☐ + 3 = 10

c. ☐ + 2 = 10

d. ☐ + 1 = 10

e. ☐ + 0 = 10

⚠ Problems

Brain-teaser Sasha built a tower with 8 blocks.
Sam put two more blocks on to the tower.
How many blocks are on the tower now? ☐

Brain-buster There were 10 people on the bus.
3 people got off.
How many people were still on the bus? ☐

Tips 💡
Can you find all the different ways to make 10? Use counters to help you.

🔄 Recap

We use these symbols to write number facts.

+   This symbol means **add** or **plus**.

=   This symbol mean **is equal to**.

2 ladybirds plus 3 woodlice is equal to 5 bugs.

2 + 3 = 5

📄 Learn

There are 5 green cubes and 3 blue cubes.
There are 8 cubes altogether.
5 plus 3 is equal to 8.
5 + 3 = 8

5 is a part.
3 is a part.
8 is the whole.

There are 4 red counters and 5 blue counters.
There are 9 counters altogether.
4 and 5 is equal to 9.
4 + 5 = 9

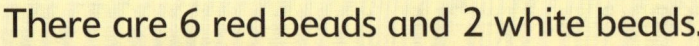

There are 6 red beads and 2 white beads.
There are 8 beads altogether.
6 plus 2 is equal to 8.
6 + 2 = 8

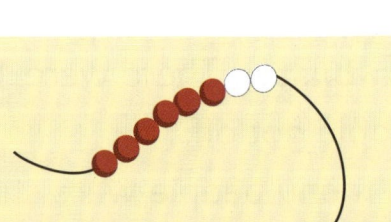

💬 Talk maths

Roll two dice. Add the numbers together.
Say the number sentence aloud.

4 plus 3 is equal to 7.

✓ Check

1. Complete the number sentence to go with each tower.

   a. 2 + 7 = ☐

   b. 3 + ☐ = ☐

   c. ☐ + ☐ = ☐

2. Use two different colours.
   Colour in the cubes to match
   the number sentence.   1 + 5 = ☐

   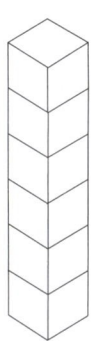

⚠ Problems

Brain-teaser: First there was one bird sitting on the tree.
Then 5 more sat on the tree.
Now, how many birds are sitting on the tree? ☐

↻ Recap

We use these symbols to write number facts.

- This symbol means **subtract** or **take away**.
= This symbol means **is equal to**.

5 – 2 = 3

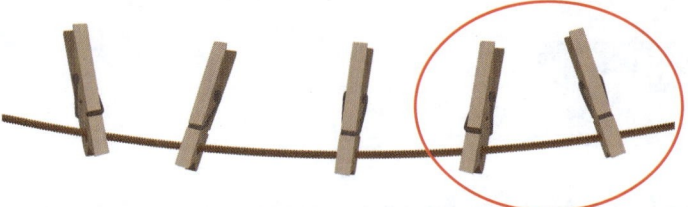

📄 Learn

First, there are 8 counters. Then we take away 3 counters. Now there are 5 counters.

8 – 3 = 5

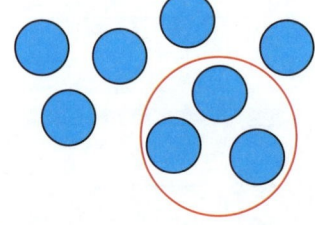

First there are 9 beads.
Then we take away 7 beads.
Now there are 2 beads.

9 – 7 = 2

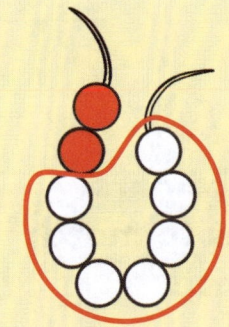

Start at number 7. Jump back 3. We land on 4.
7 – 3 = 4

Start from the first number and count back.

Talk maths

9 6 3

What **subtraction** number sentences can you make using the number cards?

Say the number sentences aloud.

☐ − ☐ = ☐

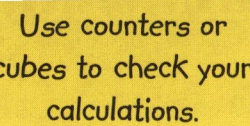

Use counters or cubes to check your calculations.

Can you spot any patterns?

✓ Check

1. Write the missing numbers.

   a. 8 − 2 = ☐   b. 8 − 3 = ☐   c. 8 − 4 = ☐

   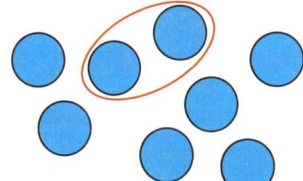 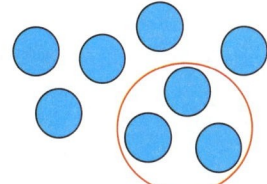 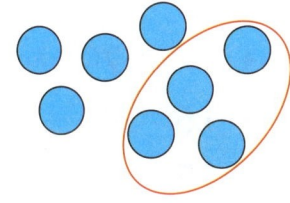

2. Write the missing numbers.

   a. 9 − 5 = ☐   b. 8 − 5 = ☐   c. 7 − 5 = ☐

   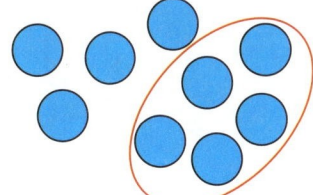 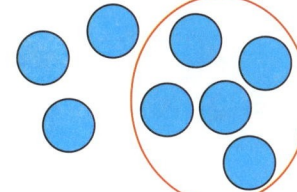 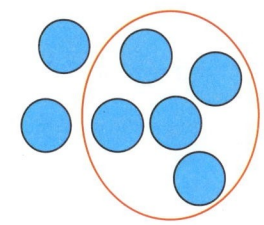

⚠ Problems

Brain-teaser: There were 7 children playing football. 2 left.

How many children are playing football now? ☐

83

⟲ Recap

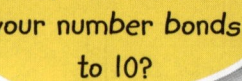

Number bonds are the **addition** and subtraction facts for a number. For example, number bonds to 10 are the addition and subtraction facts to 10.

0 + 10 = 10   10 − 10 = 0

10 + 0 = 10   10 − 0 = 10

📄 Learn

Here are the number bonds to 20:

0 + 20 = 20	10 + 10 = 20	20 − 20 = 0	20 − 10 = 10
1 + 19 = 20	11 + 9 = 20	20 − 19 = 1	20 − 9 = 11
2 + 18 = 20	12 + 8 = 20	20 − 18 = 2	20 − 8 = 12
3 + 17 = 20	13 + 7 = 20	20 − 17 = 3	20 − 7 = 13
4 + 16 = 20	14 + 6 = 20	20 − 16 = 4	20 − 6 = 14
5 + 15 = 20	15 + 5 = 20	20 − 15 = 5	20 − 5 = 15
6 + 14 = 20	16 + 4 = 20	20 − 14 = 6	20 − 4 = 16
7 + 13 = 20	17 + 3 = 20	20 − 13 = 7	20 − 3 = 17
8 + 12 = 20	18 + 2 = 20	20 − 12 = 8	20 − 2 = 18
9 + 11 = 20	19 + 1 = 20	20 − 11 = 9	20 − 1 = 19
	20 + 0 = 20		20 − 0 = 20

You can use counters to help you.

11 + 9 = 20   9 + 11 = 20

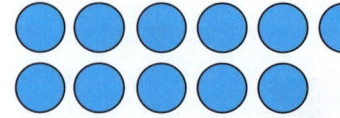 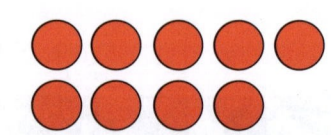

You can also use a number line.

18 + 2 = 20

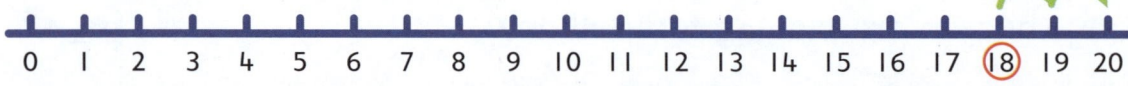

Talk maths

Learn your number bonds to 20 off by heart.

What is the same about these two number facts? Tell a friend or an adult.

17 + 3 = 20   3 + 17 = 20

20 − 5 = 15   20 − 15 = 5

Explain these facts to a friend or an adult.

15 + 5 = 20   5 + 15 = 20

12 + 8 = 20   8 + 12 = 20

20 − 2 = 18   20 − 2 = 18

Can you spot any patterns?

✓ Check

1. Write the missing numbers.
   a. 11 + ☐ = 20
   b. 12 + ☐ = 20
   c. 13 + ☐ = 20
   d. 14 + ☐ = 20
   e. 15 + ☐ = 20

2. Write the missing numbers.
   a. 20 − 14 = ☐
   b. 20 − 13 = ☐
   c. 20 − 12 = ☐
   d. 20 − ☐ = 14
   e. 20 − ☐ = 13

⚠ Problems

Brain-teaser: Sasha picks 16 daisies and 4 buttercups. How many flowers does she have altogether?

☐ + ☐ = ☐

↻ Recap

Other words we use when we talk about adding include: altogether, and, put together and total.

+ This symbol means **add** or **plus**.
= This symbol means **is equal to**.

📝 Learn

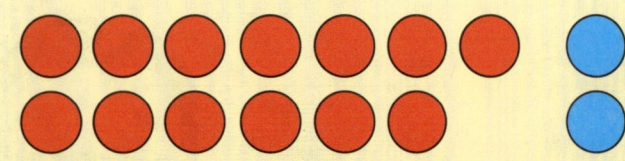

There are 13 red counters and 2 blue counters. There are 15 counters altogether.

13 + 2 = 15 or 2 + 13 = 15

There are 16 red beads and 3 white beads.
There are 19 beads altogether.

16 + 3 = 19 or 3 + 16 = 19

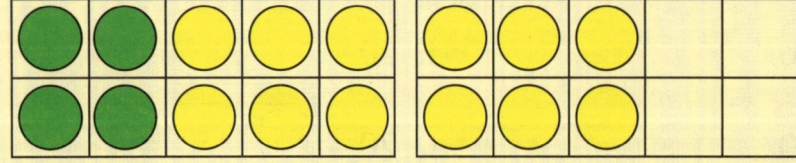

There are 4 green counters and 12 yellow counters.
There are 16 counters altogether.

Use a number line to help you add numbers together.

11 + 7 = ☐

Start with 11. Count on 7.

11 + 7 = 18 or 7 + 11 = 18

💬 Talk maths

Use the numbers on the cards. Choose two numbers. Add them together.

| 12 | 3 | 6 | 8 |

How many number sentences can you make? Say the number sentences aloud.

☐ + ☐ = ☐

✓ Check

1. **Draw lines to match the number sentences with the same answers.**

   One has been done for you.

11 + 7		6 + 10
18 + 1		13 + 5
5 + 15		12 + 7
12 + 3		8 + 7
14 + 2		17 + 3

   Start with the bigger number and count on.

⚠ Problems

Brain-teaser: Jonas and Lara went on a bug hunt. They found 6 caterpillars and 7 beetles.

How many bugs did they find altogether? ☐

Brain-buster: Pia and Alec went on a bug hunt. Pia found 8 bugs but Alec found 0.

How many bugs did they find altogether? ☐

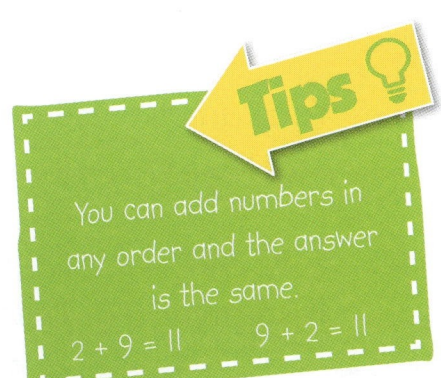

Tips: You can add numbers in any order and the answer is the same.
2 + 9 = 11   9 + 2 = 11

Subtracting numbers to 20

↻ Recap

– This symbol means **subtract** or **take away**.

= This symbol means **is equal to**.

Other words we use when talking about subtracting include: distance between and difference between.

📄 Learn

Use counters to help you subtract.

12 – 4 = ☐

Start with 12 counters and take away 4.
There are 8 counters left.

12 – 4 = 8

You can also subtract using a number line.

14 – 6 = ☐

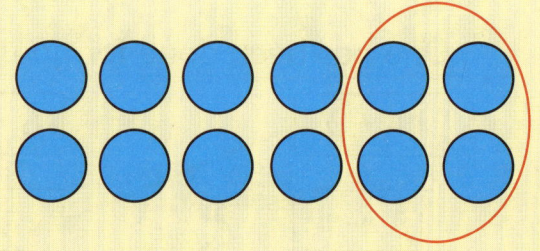

Start at 14 and count backwards 6.

14 – 6 = 8

You can also use partitioning.

17 – 12 = ☐

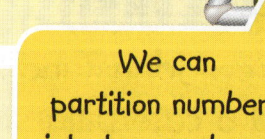

We can partition numbers into tens and ones.

Start with 17. 12 can be partitioned into ten and two.
Take away 10. Then take away 2.

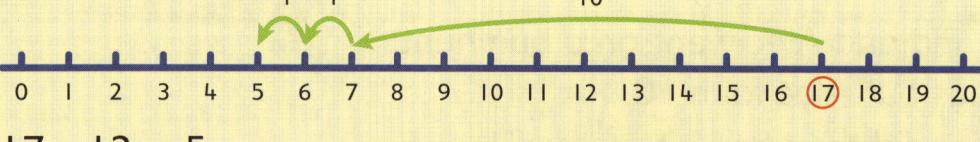

17 – 12 = 5

💬 Talk maths

Solve these number sentences.

Use your finger to jump backwards along the number line.

Say the number sentences aloud.

12 – 3 = ☐   14 – 1 = ☐   15 – 8 = ☐

✓ Check

1. Write the missing numbers.

   a. 18 – 3 = ☐   b. 16 – 5 = ☐

   c. 14 – 2 = ☐   d. 15 – 6 = ☐

   e. 13 – 7 = ☐   f. 18 – 12 = ☐

   g. 17 – 0 = ☐   h. 14 – 0 = ☐

⚠ Problems

Brain-teaser: There were 12 strawberries on a plant. Ava picked 7.

How many strawberries were left? ☐

Brain-buster: A baker made 18 cupcakes. She sold 15.

How many cupcakes were left? ☐

Problem solving

Recap

We can use number facts we know to find other facts.

2 + 5 = 7

So we also know that:

5 + 2 = 7   7 − 5 = 2   7 − 2 = 5

Learn

Look at these problems.

Find the missing number

6 + ☐ = 9

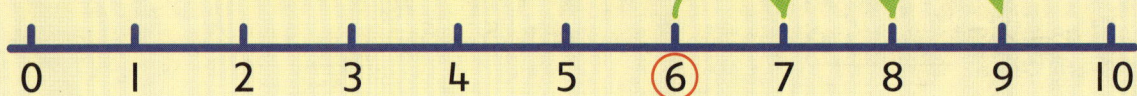

Start at 6. Count on to 9. How many did you add?

6 + 3 = 9   The missing number is 3.

Find the missing number

13 − ☐ = 8

Start with 13 counters. How many do you need to take away to leave 8?

13 − 5 = 8   The missing number is 5.

The gardener planted 12 seeds. 8 plants grew. How many plants did not grow?

12 − 8 = 4

4 seeds did not grow.

💬 Talk maths

Discuss with a friend or adult how you can solve this problem.

Jun had 16 sweets. He gave 3 to his friend. How many sweets did Jun have left?

✓ Check

1. Find the missing number.

   a. ☐ + 3 = 8

   b. 5 + ☐ = 10

   c. ☐ + 6 = 13

   d. 9 − ☐ = 7

   e. ☐ − 3 = 6

   f. 14 − ☐ = 9

⚠ Problems

Use number facts you have learned to help you.

Brain-teaser: 8 children were on the football team. 3 more children joined the team.

How many children are on the football team now? ☐

Brain-buster: Ben had 14 marbles. Pia had 5 less marbles than Ben.

How many marbles did Pia have? ☐

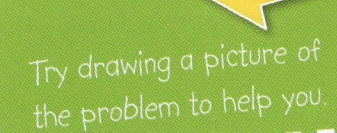

Try drawing a picture of the problem to help you.

Grouping

Here are some doubles facts:
1 + 1 = 2
2 + 2 = 4
3 + 3 = 6
4 + 4 = 8
5 + 5 = 10

↺ Recap

We can double numbers.

3 + 3 = 6
There are 2 groups of 3.

🗒 Learn

10 + 10 = 20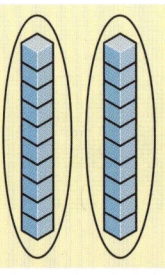
There are 2 groups of 10.
20 cubes altogether.

4 + 4 = 8
There are 2 rows of 4.
There are 8 eggs altogether.

5 + 5 + 5 = 15
There are 3 groups of 5.
15 pieces of pizza altogether.

2 + 2 + 2 = 6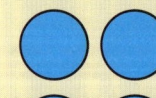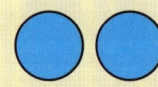
This array has 3 rows. Each row has 2 spots.
There are 6 spots altogether.

Talk maths

Count aloud in 2s, 5s, and 10s.

0, 2, 4, 6, 8, 10, 12, 14, 16, 18, 20
0, 5, 10, 15, 20, 25, 30, 35, 40, 45, 50
0, 10, 20, 30, 40, 50, 60, 70, 80, 90, 100

Check

1. Draw lines to match each array to the correct number sentence.

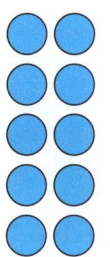 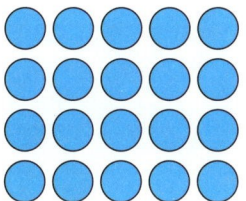 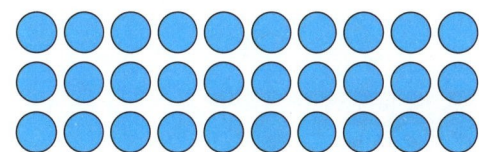

| 3 lots of 10 = 30 | 5 groups of 2 = 10 | 4 rows of 5 = 20 |

Can you count in 5s to solve this problem?

Problems

Brain-teaser: The gardener had 6 pots. She planted 2 flowers in each pot. How many flowers were there altogether? ☐

Tips: Practise counting in steps of 2, 5 and 10.

Brain-buster: The teacher had 5 pots. He put 5 pencils in each pot. How many pencils were there altogether? ☐

Sharing

↻ Recap

We can share items equally.

There were 6 toy cars.
Marc and Alec had 3 cars each.

The word **equally** means there is the same amount in each group.

📋 Learn

14 **shared by** 2 is equal to 7.

12 shared by 2 is equal to 6.

15 shared by 5 is equal to 3.

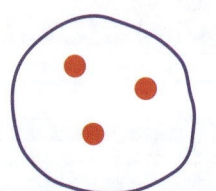 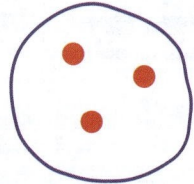 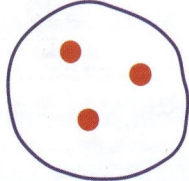 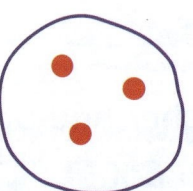 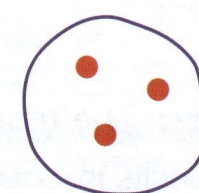

You can use objects to help you.

You can draw a picture to help you.

💬 Talk maths

How could you solve this problem? Discuss your ideas with a friend or an adult.

Try counting in 2s aloud!

There were 18 strawberries and 2 children. The strawberries are shared equally.

How many strawberries do the children have each?

✓ Check

1. Write the missing numbers.

a.  4 shared between 2 = ☐   b.  8 shared between 2 = ☐

c.  10 shared between 2 = ☐   d.  12 shared between 2 = ☐

e.  14 shared between 2 = ☐   f.  16 shared between 2 = ☐

⚠ Problems

Brain-teaser: There are 10 children and 2 tables. Each table has an equal number of children.

How many children are on each table? ☐

Brain-buster: They are 20 pencils and 5 pots. The pencils are shared equally.

How many pencils are in each pot? ☐

Tips 💡

Remember to share equally.

Halves

↻ Recap

We can share items equally into 2 groups.

8 shared by 2 is equal to 4.

📄 Learn

When you share something into 2 equal groups, you have 2 parts of a whole.

One of the two parts is called a **half**.

A half is 1 of 2 equal parts.

You can find half of an object.

You can find half of a shape.

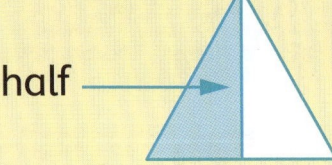 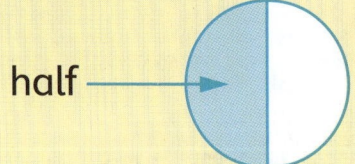

You can find half of an amount.

Half of 6 = 3

You can share 6 into 2 equal groups.

💬 Talk maths

Which shapes are divided in half? Talk to a friend or adult about how you know.

 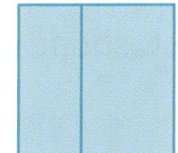  

✓ Check

1. Colour in half of each shape.

   a. 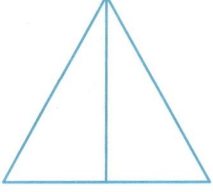   b. 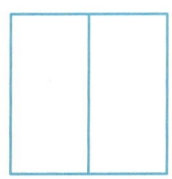   c.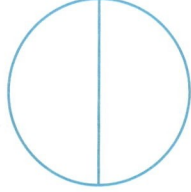

2. Find half of each amount.

   a. half of 8 = ☐

   b. half of 10 = ☐

   c. half of 12 = ☐

⚠ Problems

Brain-teaser: Ella and Maya share a packet of stickers equally. There are 10 stickers.

How many stickers do they get each?

Two halves make a whole.

Quarters

🔄 Recap

A half is one of two equal parts of an object, shape or amount.

📄 Learn

When you share something into 4 equal groups, you have 4 parts of a whole.

One of the four parts is called a **quarter**.

A quarter is 1 of 4 equal parts.

You can find a quarter of an object.

← quarter

You can find a quarter of a shape.

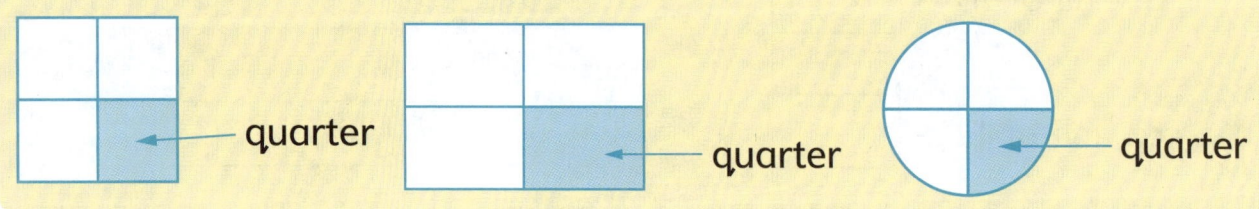

You can find a quarter of an amount.

quarter of 8 = 2   You can share 8 into 4 equal groups.

Talk maths

Which shapes are divided into quarters? Talk to a friend or adult about how you know.

  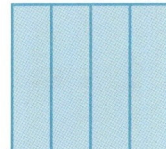 

Check

1. **Colour in a quarter of each shape.**

   a.    b. 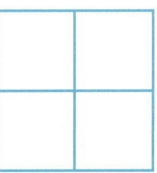   c.

2. **Find a quarter of each amount**

   a. quarter of 8 = ☐

   b. quarter of 12 = ☐

   c. quarter of 16 = ☐

   Use counters to help you. Count out the whole amount. Share the whole amount into four equal groups. One of the groups is equal to a quarter.

Problems

Brain-teaser: Lara had 20 sweets. She shared them equally between her four friends.

How many did they have each? ☐

Brain-buster: What is one quarter of 20? ☐

Tips: Four quarters make a whole.

Length

↻ Recap

We can compare heights.

Max is taller than Maya.
Maya is shorter than Max.

Max   Maya

📄 Learn

Use the words **longer** and **shorter** to compare lengths.

The spotty snake is longer than the stripy snake.
The stripy snake is shorter than the spotty snake.

You can measure how long something is using counters.

If you don't have any counters you can use other items such as pasta, buttons, toy bricks, stones or leaves.

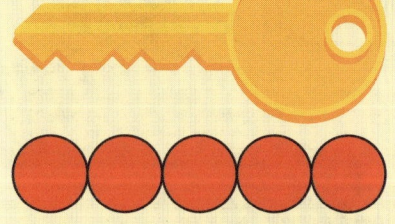

The key is 5 counters long.

You can also use a ruler to measure items.
We use **centimetres** to measure length.
You can write centimetres as cm.

The pencil is 12cm long.

💬 Talk maths

Can you say these words we use for measuring length aloud?

centimetre   taller   shorter   longer

Choose one of the words. Think of a sentence using that word.

✓ Check

1. Read the ruler. Where is the arrow pointing?

   a.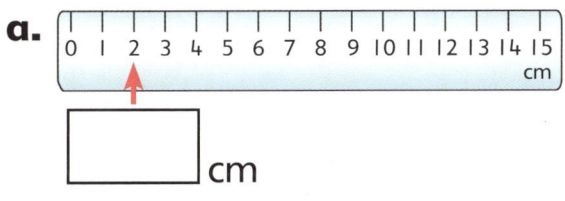
   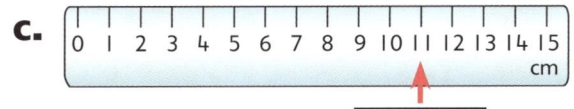 cm

   b.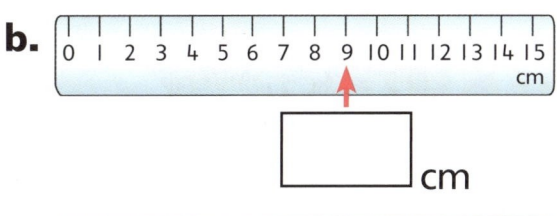
   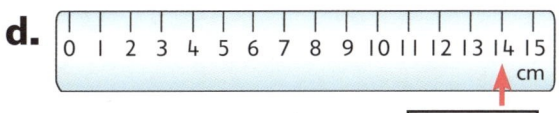 cm

   c.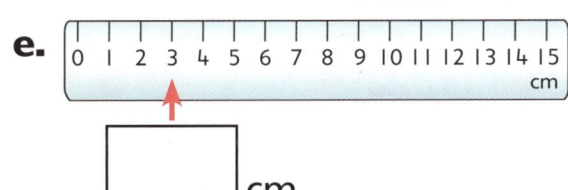
   ▢ cm

   d.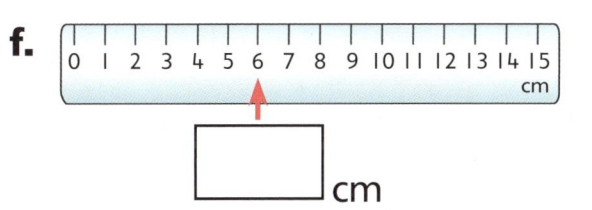
   ▢ cm

   e. ▢ cm

   f. ▢ cm

⚠ Problems

Brain-teaser: How many counters long is the pencil?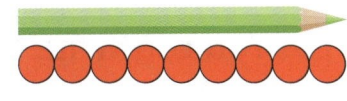

The pencil is ▢ counters long.

Brain-buster: Max is taller than Maya. Beth is shorter than Max but taller than Maya. Who is the shortest?

_____ is the shortest.

💡 Tips

When measuring how long or tall something is, make sure the ruler is flat. Make sure the 0 is at the beginning of the item you are measuring.

Can you measure your pencil? How many counters is it? How many cm is it?

Mass and weight

"The words **light** and **heavy** describe the weight of an item."

↻ Recap

The feather is light.   The brick is heavy.

📄 Learn

Use the words **lighter** and **heavier** to compare how heavy items are.

The feather is lighter than the brick.
The brick is heavier than the feather.

You can weigh items using blocks.
The pencil weighs 3 blocks.

You can use a scale to measure items.

We use **grams** to measure weight.

You can write grams as g.

The paintbrush is 16g.

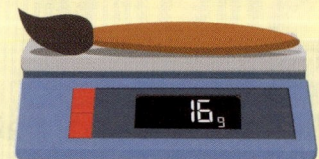

💡 Tips

You can write grams as g.

🗨 Talk maths

There are lots of different types of weighing scales.

Discuss the different types of scales with a friend or an adult.

Where might you find these weighing scales?

What would you weigh on them?

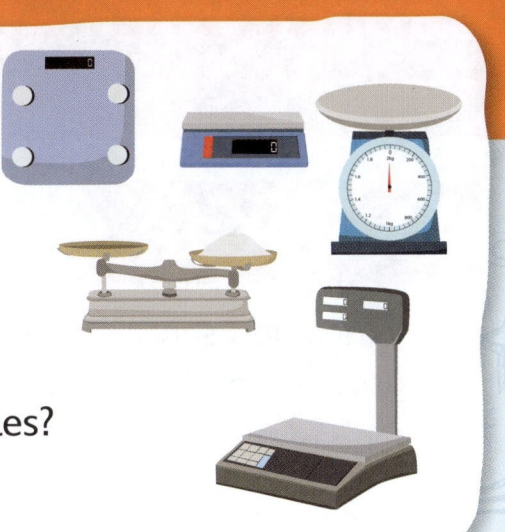

✓ Check

1. How many blocks does each item weigh?

   a.  ☐ blocks   b.  ☐ blocks

   c.  ☐ blocks

2. How much does the heaviest parcel weigh? ☐ g

⚠ Problems

Brain-teaser: A pencil weighs 3 blocks.

A glue stick weighs 9 blocks.

How many blocks do the pencil and glue stick weigh altogether?

☐ blocks

Capacity and volume

A half is 1 of 2 equal parts.

↻ Recap

When you share something into two equal groups, you have two parts of a whole.

One of the two parts is called a **half**.

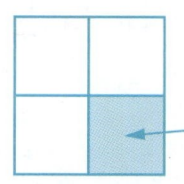 ← half

When you share something into four equal groups, you have four parts of a whole.

One of the four parts is called a **quarter**.

A quarter is 1 out of 4 equal parts.

quarter

📄 Learn

Use the words **full** and **empty** to describe how much liquid there is in a container.

The watering can is full.

The watering can is empty.

A container might also be **half full** or a **quarter full**.

The bucket is half full.

The measuring cylinder is a quarter full.

💬 Talk maths

There are lots of different types of container.
Where might you find these containers?
Which container do you think holds the most water? Explain your answer to a friend or adult.

💡 Tips
Look for other containers for holding water.

✓ Check

1. Draw lines to match each jug to the correct description.

| empty | half full | quarter full | full |

⚠ Problems

Brain-teaser: Pia had a full glass of water. She drank half.
How much water does she have left? Circle the correct answer.

a full glass   half a glass   quarter of a glass   an empty glass

Measurement

Time

↻ Recap

There are lots of different types of clock. We use clocks to tell the time.

📄 Learn

These are the days of the week:
Monday, Tuesday, Wednesday, Thursday, Friday, Saturday, Sunday.

These are the months of the year:
January, February, March, April, May, June, July, August, September, October, November, December.

💡 **Tips**
The days of the week and the months of the year are always in the same order.

A clock has a minute hand (longer hand) and an hour hand (shorter hand).

→ minute hand
→ hour hand

The minute hand is long. The hour hand is short.

On the hour, the minute hand points to 12.

→ The minute hand is pointing to 12
→ The hour hand is pointing to 5

5 o'clock

At half past the hour, the minute hand points to 6.

→ The hour hand is pointing halfway between 3 and 4.
→ The minute hand is pointing to 6

half past 3

The 6 is halfway round the clock face.

106

🗨 Talk maths

Here are some words we use to talk about time. Say them aloud.

| today | yesterday | tomorrow | morning |
| afternoon | evening | quicker | slower |

The tortoise was slower than the hare.

Choose one of the words. Think of a sentence using that word.

✓ Check

Tips: There are 7 days in a week. There are 24 hours in a day.

1. **Write the answers.**

   a. Fill in the missing days of the week:

   Monday, T_____, Wednesday, Thursday, F_____, Saturday, S_____

   b. Fill in the missing months of the year:

   January, February, M_____, April, May, June, J_____, August, September, O_____, November, December.

2. **Match the clock to the correct time.**

| 2 o'clock | 6 o'clock | 8 o'clock | half past 7 | half past 1 |

⚠ Problems

Brain-teaser: It is half past 10. Ava started school 2 hours ago. What time did Ava start school?

_____

Money

Recap

We use money to buy items. Many people keep their money in a purse or wallet.

Learn

Here are the coins we use:

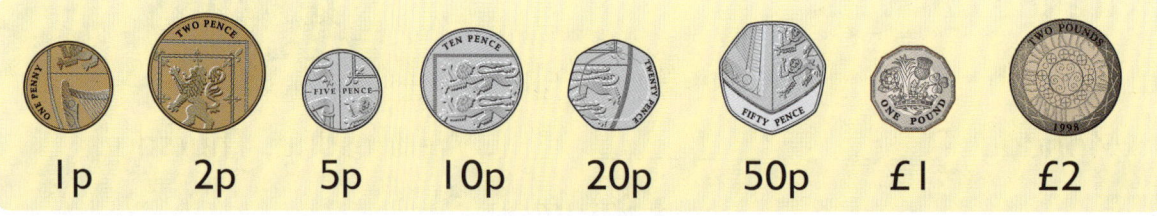

Different items cost different amounts.

To write an amount of money less than one **pound** we use the p sign for **pence**.

This pencil costs 5 pence.

We can buy the pencil using these coins:

108

💬 Talk maths

What coins could you use to buy the paintbrush?
Tell a friend or an adult.

What coins could you use to make these amounts?

7p   23p   60p

Measurement

✓ Check

1. Match each coin to its name.

    10p

   50p

    20p

   1p

    5p

   2p

⚠ Problems

Brain-teaser: How many different ways can you make 6p using coins?

Recap

2D shapes are flat.

Here are some **2D shapes**.

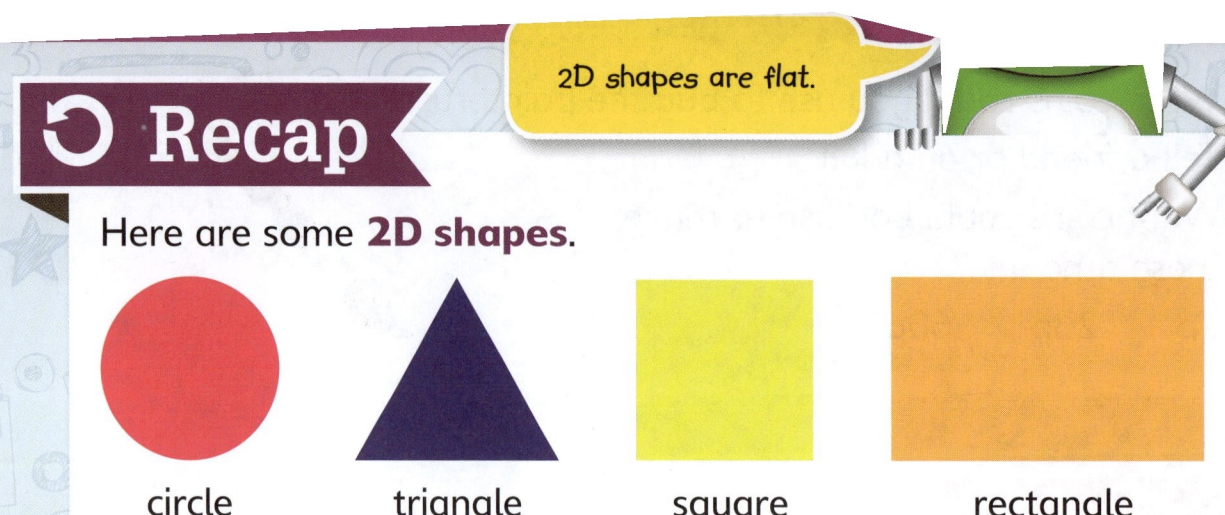

circle	triangle	square	rectangle

Learn

Not all shapes look identical.
Circle all of the triangles.

All triangles have 3 sides and 3 corners.

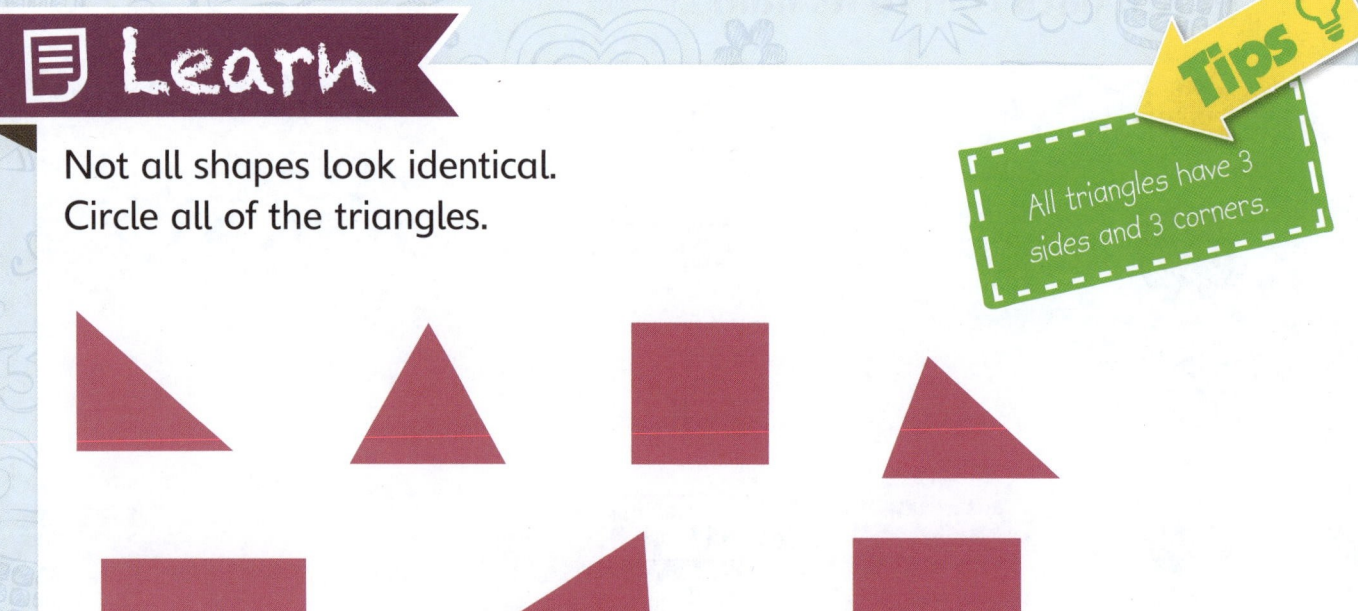

You can see shapes everywhere.

 This window is a square.

 This clock is a circle.

 This sign is a triangle.

 This phone is a rectangle.

Talk maths

Look at each shape. Count how many sides it has.

Tell a friend or an adult the name of each shape and how many sides it has.

How is the circle different from the other shapes?

✓ Check

1. Draw lines to match each shape to its name.

   rectangle   circle   triangle   square

2. Colour in the rectangles.

      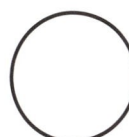 

3. Draw a circle around the circles.

⚠ Problems

Brain-teaser: Ben is thinking of a shape. What shape is Ben thinking of?

I am thinking of a shape. It has 3 straight sides.

↻ Recap

2D shapes are flat.

These are 2D shapes.

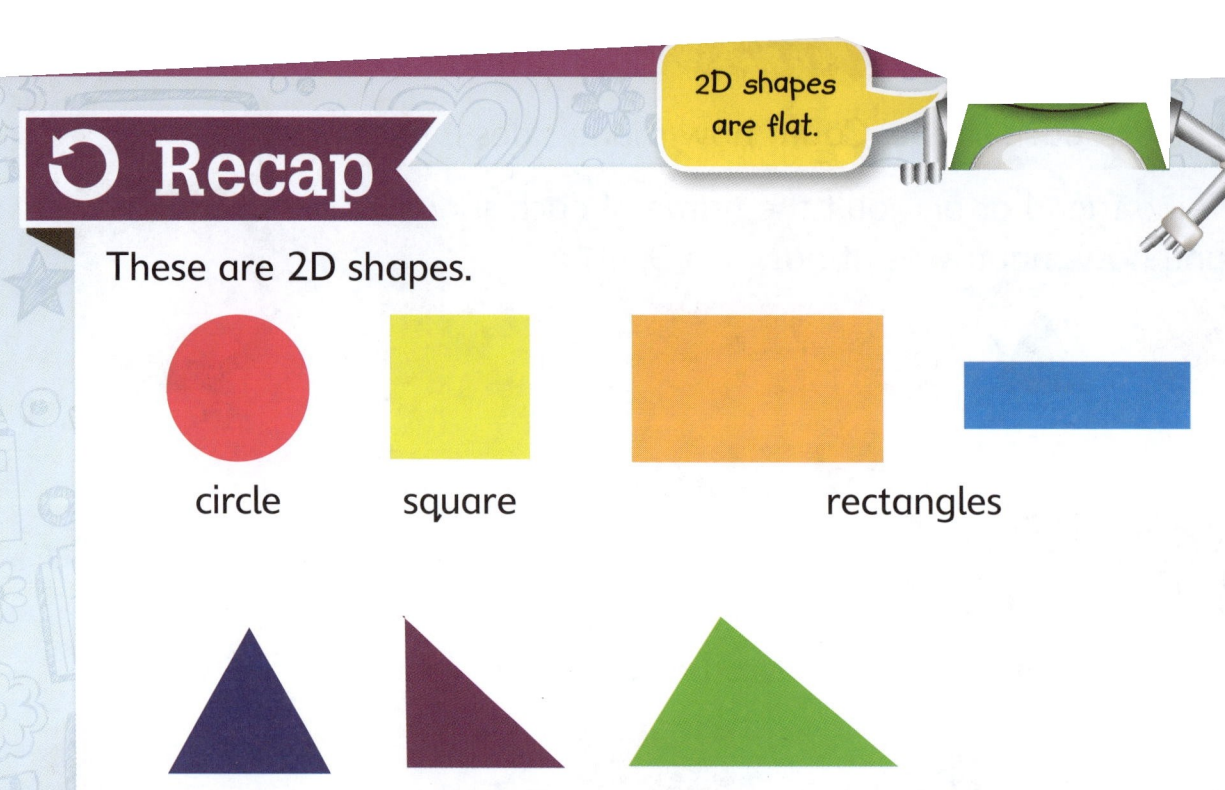

📄 Learn

3D shapes are solid.

Here are some **3D shapes**:

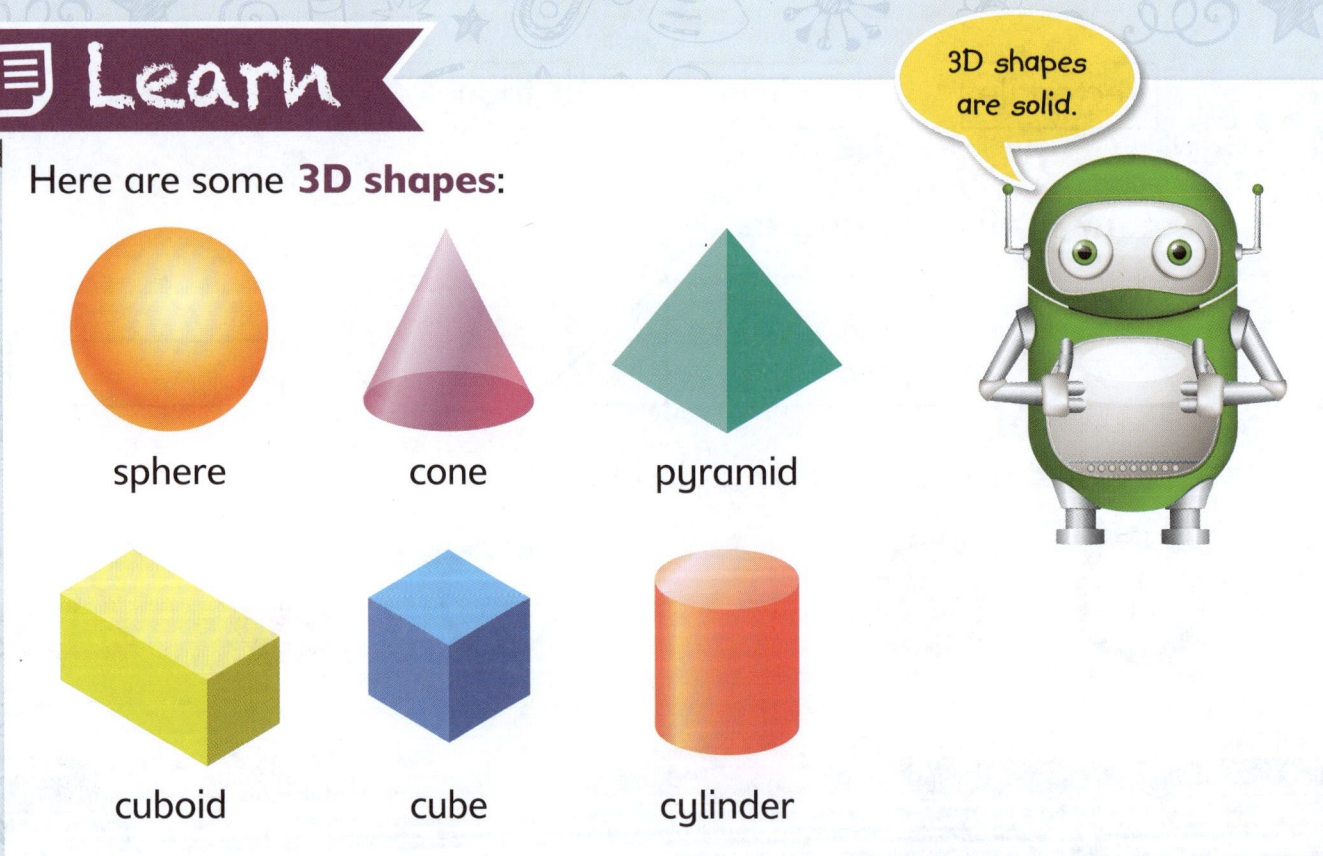

💬 Talk maths

Here are some 3D shapes in real life.

Work with a friend or an adult. What is each shape called?

✓ Check

1. Draw a circle around the 3D shapes.

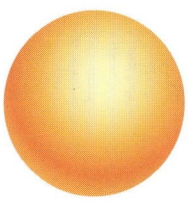  

sphere　　　　　pyramid　　　　　triangle

circle　　　　　cuboid　　　　　square

⚠ Problems

Brain-teaser: What is this shape called? Draw a line to the correct shape word.

cuboid

sphere

pyramid

Recap

> Sam is behind the desk.
> The cat is under the desk.
> The books are on top of the desk.

We use these words to describe the position of something:
above, below, behind, in front, on top, left, right, in the middle, underneath, inside, outside.

Learn

Ava is sitting in front of Ben.

Max is sitting behind Sam.

Ben is sitting between Ella and Max.

Describe to an adult or friend where Alec is sitting.

Tips

When learning your left and right, remember which hand you write with – is it your left hand or your right hand?
Or hold your hands up in front of you and move your thumbs to make an 'L' shape – the hand that makes an L is your left hand.

🗨 Talk maths

To describe the direction something moves, use the words: forwards, backwards, left and right.

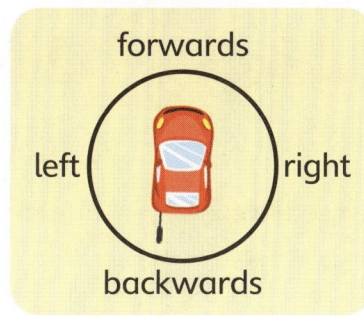

Amira can move the car forwards, backwards, left and right.

Use these turns to talk about how the car can change direction.

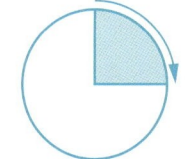 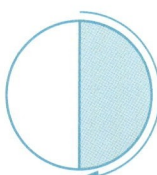 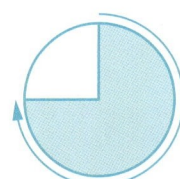 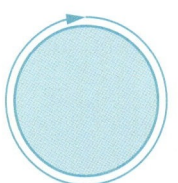

Quarter turn   Half turn   Three-quarter turn   Whole turn

✓ Check

1. What will the car be facing if it makes…

   a. a half turn?
   _____

   b. a whole turn?
   _____

   c. a quarter turn to the left?
   _____

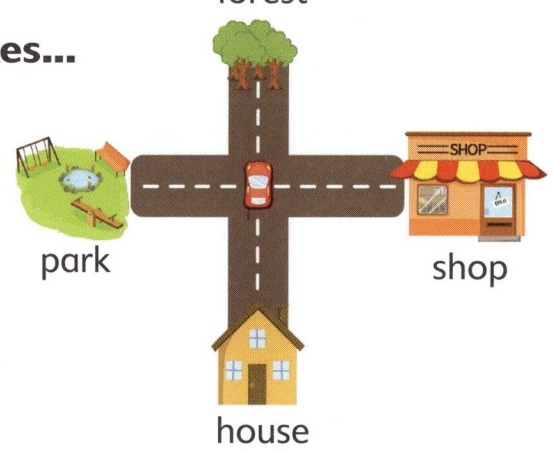

⚠ Problems

Brain-teaser: A robot is facing the front of the classroom. It makes a whole turn.

Where is the robot facing now? Tick the correct answer.

☐ the back of the classroom   ☐ the front of the classroom

English glossary

A
apostrophes look like this ' and can be used to show missing letters in a word.

B
blending is when we join the sounds together to read a word.

C
capital letters are A, B, C, D, E, F, G, H, I, J, K, L, M, N, O, P, Q, R, S, T, U, V, W, X, Y and Z.

characters are the people or animals in a story.

compound words are words made from two or more shorter words. Some compound words are toothpaste and sunshine.

contractions are shortened words, where an apostrophe is used to replace missing letters. Contractions include I'll and we've.

E
events are the things that happen in a story.

exclamation marks look like this ! and are punctuation marks. Exclamation marks are used at the end of a sentence to emphasise something, to show shock or surprise, or when someone is shouting.

F
full stops look like this . and are punctuation marks. We use full stops at the end of a sentence.

I
inference is when we use clues to work something out when it is not explicit in the text.

J
joining words join two words or groups of words together. Joining words include and.

P
past tense tells us that something has already happened.

plural means more than one.

predict means to guess what will happen before it happens.

prefix is a letter or group of letters that is added to the start of a word to change the meaning.

Q
question marks look like this ? and are punctuation marks used at the end of a question or when you are asking something.

S
segmenting is when we split a word up into its sounds to help us spell the word.

sentences are groups of words that make sense.

singular means there is only one.

suffix is a letter or group of letters added to the end of a word to change its meaning.

T
titles are how we name a piece of writing. A text might have a heading like How cars work. A story might have a name like The Yellow Balloon.

Maths glossary

2D shapes Two dimensional shapes are flat. Here are some examples:

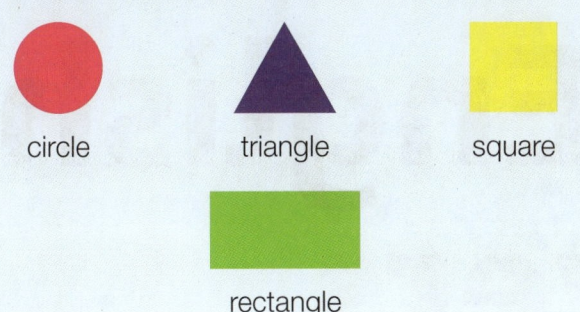

3D shapes Three dimensional shapes are solid shapes. Here are some examples:

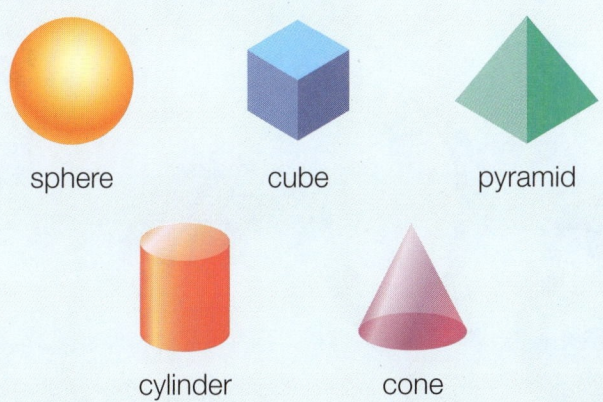

A
Addition When two or more things are added together. + This is the addition symbol.

C
Centimetres A measure of length or height. Centimetres can be written as cm.

G
Grams A measurement of mass or weight. Grams can be written as g.

H
Half When something is split equally into two parts, one of the parts is called half.

N
Number bonds The addition and subtraction facts for a number. For example, number bonds to 10 are the addition and subtraction facts to 10.

O
Ordinal numbers 1st, 2nd, 3rd, 4th, 5th, 6th, 7th, 8th, 9th, 10th are ordinal numbers.

P
Pence An amount of money. You can write pence as p, for example, 1p.

Pound An amount of money. The symbol for pound is £.

S
Subtraction When something is taken away. – This is the subtraction symbol.

Q
Quarter When something is split into four equal parts, one part is called a quarter.

Maths & English Answers

English answers

SENTENCES

Page 11

1. Ⓐ a Ⓝ e Ⓔ g r Ⓠ
   d Ⓖ Ⓓ Ⓑ b Ⓡ i Ⓜ

2. n → N, c → C, g → G, h → H, j → J, t → T

3. a. **T**he sun is shining**.**
   b. **I** went to the park on Saturday**.**
   c. **I** have a pet zebra and a whale lives in my bath**.**

4. a. **W**e had pizza for lunch**.**
   b. **M**y coat is red**.**
   c. **T**he teacher told us a joke**.**

Page 13

2. I like chocolate ice cream. ☐
   Can an elephant run? ✓
   My favourite animal is an elephant. ☐
   How do you make ice cream? ✓

3. a. Why is a cactus pricky**?**
   b. Did dinosaurs have feathers**?**
   c. How many stripes does a zebra have**?**

4. Accept any question written which starts with a capital letter and ends with a question mark.

Page 15

2. a. Help, I'm melting**!**
   b. That is amazing**!**
   c. We are going on holiday**!**
   d. My brother ate broccoli**!**
   e. The headteacher did a headstand**!**
   f. I found the missing puzzle piece**!**
   g. I cannot wait to go swimming**!**

Page 16–17

1.
Capital letters	A	B	C	D	E	F	G
Capital letters	H	I	J	K	L	M	N
Capital letters	O	P	Q	R	S	T	U
Capital letters	V	W	X	Y	Z		

2. a. (Ava) went to the park.
   b. (Mr Green) was a teacher.
   c. Yesterday, (Max) made a cake.
   d. The lady who lives next door is called (Mrs Jackson).
   e. (Ben) went to the park with (Jun).
   f. (Alec) and (Tom) had ice cream after lunch.

3. a. **I** like hopping like a grasshopper.
   b. When it started raining, **I** splashed in the puddles.

Page 18–19

1. a. Foxes hunt at night. = 3 claps
   b. Bees make honey. = 2 claps
   c. Plums are purple. = 2 claps

2. b. Giraffes | have | blue | tongues.
   c. The | snail | slithered | down | the | path.

3. a. Owls have big eyes.
   b. There was a frog in the pond.
   c. Some dinosaurs fly.

4. a. I | like | pizza.
   b. I like pizza.

Page 21

1. a. I like sweets.
   b. The lion roared loudly.
   c. The rocket went into space.
   d. The hippo rolled in mud.

2. a. sky. stars the shining are in The ✗
   b. Ben rode his bike at the park. ✓
   c. Cheetahs have spots. ✓
   d. legs. have Snakes do not ✗

Page 23

1. a. My bike is blue **and** Tom's bike is red.
   b. Pia is playing with the bricks **and** Jun is reading a book.
   c. Sam had an apple **and** Sasha had a banana.

2. a. Cheetah are fast **and** sloths are slow.
   b. Dogs bark **and** cats purr.
   c. Elephants live on land **and** whales live in the sea.

Page 25

1. b. Then she baked fresh bread. **2**
   The baker sold the fresh bread. **3**
   The baker got up early to bake. **1**
   c. It had snowed. **1**
   The boy put a hat and a scarf on his snowman. **3**
   The boy built a snowman. **2**
   d. The gardener planted some seeds. **1**
   The carrots grew. **2**
   The gardener ate the carrots. **3**
   e. She kicked the ball. **2**
   She scored a goal. **3**
   Ella put on her football boots. **1**

VOCABULARY

Page 27

1. a hat**s**
   b. tractor**s**
   c. tree**s**
   d. sock**s**
   e. cake**s**
   f. star**s**

2. a. box**es**
   b. wish**es**
   c. watch**es**
   d. brush**es**
   e. dress**es**

119

Page 29

1. a. look**ing**
   b. paint**ing**
   c. help**ing**
   d. sing**ing**
   e. grow**ing**

2. a. I **look** out of the window.
      I am **looking** out of the window.
   b. He is **painting** the door.
      He can **paint** doors.
   c. I am **helping** my mum to wash the car.
      I **help** Gran cook the dinner.
   d. I can **sing**.
      I like **singing**.
   e. Flowers **grow** in the garden.
      We are **growing** sunflowers.

Page 31

1. a. walk**ed**   e. jump**ed**
   b. play**ed**   f. cook**ed**
   c. help**ed**   g. wash**ed**
   d. look**ed**   h. pull**ed**

2. a. smash**ed**   f. thump**ed**
   b. bang**ed**   g. crash**ed**
   c. push**ed**   h. wish**ed**
   d. pick**ed**   i. buzz**ed**
   e. clean**ed**

Page 33

1. a. small**er**   small**est**
   b. long**er**   long**est**
   c. old**er**   old**est**
   d. new**er**   new**est**

2. a. short**er**
      short**est**
   b. high**er**
      high**est**

3.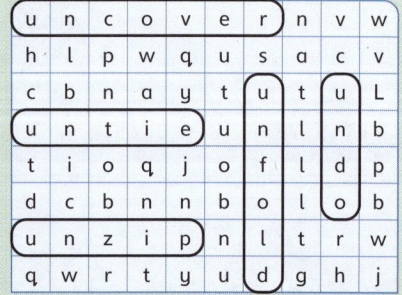

Page 35

1. a. **un**happy   e. **un**fold
   b. **un**tie   f. **un**cover
   c. **un**do   g. **un**zip
   d. **un**block   h. **un**kind

2.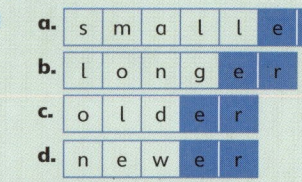

SPELLING

Page 36–37

1.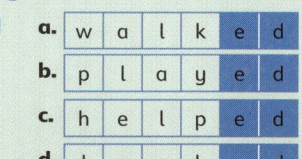

2. Monday, Tuesday, Wednesday, Thursday, Friday, Saturday, Sunday

3. M o n **d a y**
   T u e s d a y
   W e d n e s **d a y**
   T h u r s d a y
   F r i d a y
   S a t u r d a y
   S u n d a y

Page 39

1. a. thi**nk**
   b. li**nk**
   c. ta**nk**
   d. ba**nk**

2. a. sa**nk**
   b. pi**nk**
   c. dri**nk**
   d. si**nk**

Page 41

1. a. h a **tch**
   b. **ch** i p
   c. l a **tch**
   d. **ch** e s t
   e. s n a **tch**
   f. i **tch**

2. a. (match)   mach
   b. (bench)   bentch
   c. cruntch   (crunch)
   d. (pitch)   pich
   e. churtch   (church)
   f. (pinch)   pintch

Page 43

1.
h	d	s	t	b	c	v	w
a	n	l	p	j	h	t	r
v	r	e	a	b	o	v	e
e	t	e	k	w	r	t	n
n	w	v	n	v	v	c	r
l	t	e	m	g	i	v	e
m	o	v	e	k	l	n	m
p	d	t	v	a	l	v	e

2. shelve, shelv, curve, starve, starv
   activ, active, massive, nerv, weave
   weav, curv, nerve, move, massiv

3. a. glo**ve**
   b. oli**ve**
   c. lo**ve**
   d. do**ve**

Page 45

1. a. d o l **ph** i n
   b. s **ph** e r e
   c. e l e **ph** a n t
   d. **ph** o n e
   e. t r o **ph** y
   f. a l **ph** a b e t

2.
f	ph
fairy	dolphin
feet	trophy
frog	phone
fun	alphabet
first	sphere

Page 47

1. a. happ**y**
   b. cherr**y**
   c. bunn**y**
   d. lad**y**
   e. pupp**y**
   f. part**y**
   g. berr**y**
   h. famil**y**

2. happee, lady, bunny, puppe, parte
   berry, happy, famile, family, bunee

3.
h	a	p	p	y	n	y	c
g	f	g	h	w	r	t	h
w	r	t	h	l	n	m	e
w	c	v	k	a	e	r	r
r	s	b	v	d	l	l	r
b	e	r	r	y	p	p	y
t	b	p	a	r	t	y	g
f	a	m	i	l	y	n	v

Page 49

1. I would → I'd
   I am → I'm
   I will → I'll
   we will → we'll
   we have → we've
   I have → I've

2. a. **I'm** going to sing a song on the stage.
   b. **I've** got a brother.
   c. **We'll** have to walk to school.
   d. **We've** got half an hour until the next bus.
   e. **I'll** wait for you.
   f. **I'd** like a drink.

Page 51

1. tooth → paste
   snow → man
   sun → shine
   pop → corn
   straw → berry

2. a. rain**drop**
   b. snow**flake**
   c. blue**berry**
   d. door**mat**

Page 54–55

1.
s	s	o	m	e	t	h	w
c	a	t	o	d	a	y	a
h	r	k	l	h	g	d	s
o	e	s	a	i	d	w	r
o	h	o	u	s	e	y	b
l	n	o	n	e	h	o	t
m	t	h	e	r	e	u	h
w	h	e	r	e	p	n	e

3. Max lives in the **house** with a red door.
   One day a robot came to school.
   There **was** a robot on the playground.
   "I can see a robot," **said** Max.

READING

Page 57

1. a. A Visit to the Vets
   b. Pia and Nan
   c. Going to the vets, needing a check-up, not getting in the box.

2. At the Park
   Dad and Max

Page 59

1. swimming ☐   fishing ✓   running ☐
2. in the kitchen ☐   in the car ☐   in the shed ✓
3. scared ✓   excited ☐   happy ☐

Page 61

1. Ben does not want to go down the slide. ✓
   Ben likes the slide because it is fast. ☐
   Ben likes the slide because it is high. ☐

2. Example explanation: I think Ben does not want to go down the slide because it is very high and fast. Ben is not happy when he looks down. He goes up the steps slowly. Ben does not like going fast or being high up.

Maths answers

NUMBER AND PLACE VALUE

Page 65
1. a. 10, 11, 12, 13, 14, 15
   b. 22, 23, 24, 25, 26, 27
2. a. 9, 8, 7, 6, 5, 4
   b. 26, 25, 24, 23, 22, 21
3. 25

Brain-teaser: 27, 28, 29, 30, 31, 32

Page 67

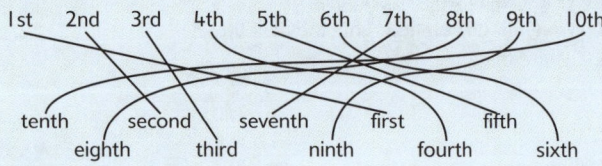

2. 1st, 2nd, 3rd, 4th, 5th

Brain-teaser: 3rd

Page 69
1. a. 0, 2, 4, 6, **8**, 10
   b. 0, 2, **4**, **6**, 8, 10
   c. 0, **2**, 4, 6, 8, **10**
   d. 8, 10, 12, **14**, 16, 18, 20
   e. 8, 10, **12**, 14, **16**, 18, 20
   f. **8**, 10, 12, 14, **18**, 20
2. 6, 8, **10**, **12**, **14**, 16, **18**, 20

Brain-teaser: 12 (2, 4, 6, 8, 10, 12)

Page 71
1. a. 0, 5, 10, 15, 20, **25**, 30
   b. 0, 5, **10**, 15, 20, 25, 30
   c. 0, 5, 10, 15, **20**, 25, 30
   d. 20, 25, **30**, 35, 40, 45, 50
   e. 20, 25, 30, 35, 40, **45**, 50
   f. 20, **25**, 30, **35**, 40, **45**, 50

Brain-teaser: 20 (5, 10, 15, 20)
Brain-buster: 30p (5, 10, 15, 20, 25, 30)

Page 73
1. a. 0, 10, **20**, 30, 40, 50,
   b. 0, 10, 20, 30, **40**, 50
   c. 0, 10, 20, 30, 40, **50**
   d. 60, **70**, 80, 90, 100
   e. 60, 70, **80**, 90, 100
   f. 60, 70, 80, **90**, 100
   g. 0, **10**, 20, **30**, 40, **50**, 60, **70**, 80, **90**, 100
   h. 0, 10, **20**, 30, **40**, 50, **60**, 70, **80**, 90, **100**

Brain-teaser: 40

Page 75
1. a. one more than 5 = 6   one less than 5 = 4
   b. one more than 15 = 16   one less than 15 = 14
   c. one more than 25 = 26   one less than 25 = 24

Brain-teaser: 16 – 1 = 15
Brain-buster: 29 + 1 = 30

Page 77

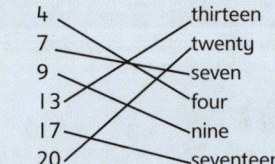

Brain-teaser: Number 16 circled

ADDITION AND SUBTRACTION

Page 79
1. a. 1 + **9** = 10
   b. 2 + **8** = 10
   c. 3 + **7** = 10
   d. 4 + **6** = 10
   e. 5 + **5** = 10
2. a. **6** + 4 = 10
   b. **7** + 3 = 10
   c. **8** + 2 = 10
   d. **9** + 1 = 10
   e. **10** + 0 = 10

Brain-teaser: 10
Brain-buster: 7

Page 81
1. a. 2 + 7 = **9**
   b. 3 + 5 = **8**
   c. 6 + 4 = **10**
      or 4 + 6 = 10
2. 1 + 5 = 6

Brain-teaser: 6

Page 83
1. a. 6
   b. 5
   c. 4
2. a. 4
   b. 3
   c. 2

Brain-teaser: 5

Page 85
1. a. 11 + **9** = 20
   b. 12 + **8** = 20
   c. 13 + **7** = 20
   d. 14 + **6** = 20
   e. 15 + **5** = 20
2. a. 20 – 14 = **6**
   b. 20 – 13 = **7**
   c. 20 – 12 = **8**
   d. 20 – 6 = **14**
   e. 20 – 7 = **13**

Brain-teaser: 16 + 4 = 20

Page 87

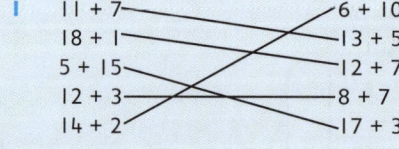

Brain-teaser: 13
Brain-buster: 8

Page 89

1.
   a. 15
   b. 11
   c. 12
   d. 9
   e. 6
   f. 6
   g. 17
   h. 14

Brain-teaser: 5
Brain-buster: 3

Page 91

1.
   a. **5** + 3 = 8
   b. 5 + **5** = 10
   c. **7** + 6 = 13
   d. 9 – **2** = 7
   e. **9** – 3 = 6
   f. 14 – **5** = 9

Brain-teaser: 11
Brain-buster: 9

MULTIPLICATION AND DIVISION

Page 93

1.
   3 lots of 10 = 30   5 groups of 2 = 10   4 rows of 5 = 20

Brain-teaser: 12
Brain-buster: 25

Page 95

1.
   a. 2
   b. 4
   c. 5
   d. 6
   e. 7
   f. 8

Brain-teaser: 5
Brain-buster: 4

FRACTIONS

Page 97

1. Half of each shape coloured.

2.
   a. 4
   b. 5
   c. 6

Brain-teaser: 5

Page 99

1. One quarter of each shape coloured.

2.
   a. 2
   b. 3
   c. 4

Brain-teaser: 5
Brain-buster: 5

MEASUREMENT

Page 101

1.
   a. 2cm
   b. 9cm
   c. 11cm
   d. 14cm
   e. 3cm
   f. 6cm

Brain-teaser: The pencil is 9 counters long.
Brain-buster: Maya is the shortest.

Page 103

1.
   a. 9
   b. 2
   c. 7

2. 90g

Brain-teaser: 12

Page 105

1.

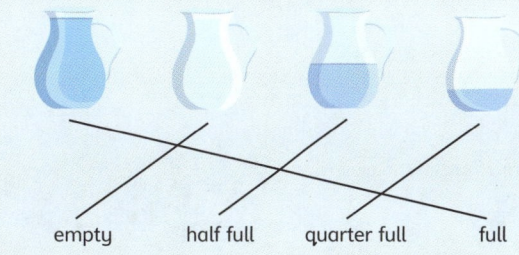

empty   half full   quarter full   full

Brain-teaser: half a glass

Page 107

1.
   a. Tuesday, Friday, Sunday
   b. March, July, October

2.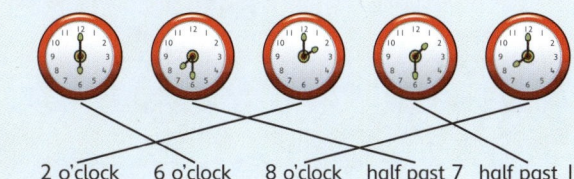

2 o'clock   6 o'clock   8 o'clock   half past 7   half past 1

Brain-teaser: half past 8

Page 109

1.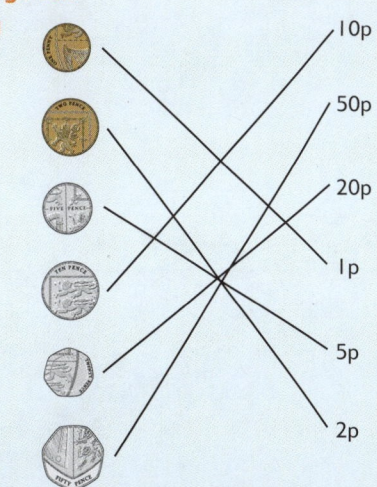

Brain-teaser: 5 different ways: 1p + 1p + 1p + 1p + 1p + 1p OR 1p + 1p + 1p + 1p + 2p OR 1p + 1p + 2p + 2p OR 2p + 2p + 2p OR 5p + 1p

GEOMETRY: PROPERTIES OF SHAPE

Page 111

1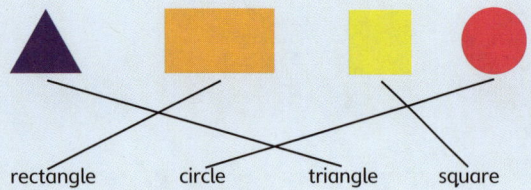

2 Both rectangles coloured
3 clock, pizza, wheel, plate circled

Brain-teaser: triangle

Page 113

1 sphere, pyramid, cuboid circled

Brain-teaser: pyramid

GEOMETRY: POSITION AND DIRECTION

Page 115

   a. house
   b. forest
   c. park

Brain-teaser: the front of the classroom

English progress tracker

Sentences

Practised Achieved

☐	☐	Capital letters and full stops	10
☐	☐	Question marks	12
☐	☐	Exclamation marks	14
☐	☐	Capital letters for names and 'I'	16
☐	☐	Spaces between words	18
☐	☐	Sequencing words	20
☐	☐	Using 'and'	22
☐	☐	Sequencing sentences	24

Reading

Practised Achieved

☐	☐	Titles, characters and events	56
☐	☐	Inference	58
☐	☐	Predict what happens next	60

Vocabulary

Practised Achieved

☐	☐	Adding 's' or 'es'	26
☐	☐	Adding 'ing'	28
☐	☐	Adding 'ed'	30
☐	☐	Adding 'er' and 'est'	32
☐	☐	Adding the prefix 'un'	34

Spelling

Practised Achieved

☐	☐	Days of the week	36
☐	☐	'nk' words	38
☐	☐	The 'ch' sound	40
☐	☐	Words ending in the 'v' sound	42
☐	☐	'ph' words	44
☐	☐	Words ending in 'y'	46
☐	☐	Contractions	48
☐	☐	Compound words	50
☐	☐	Tricky words (1)	52
☐	☐	Tricky words (2)	54

Maths progress tracker

Number and place value

Practised Achieved

☐	☐	Counting forwards and backwards	64
☐	☐	Ordinal numbers	66
☐	☐	Counting in multiples of 2	68
☐	☐	Counting in multiples of 5	70
☐	☐	Counting in multiples of 10	72
☐	☐	One more, one less	74
☐	☐	Read and write numbers to 20	76

Addition and subtraction

Practised Achieved

☐	☐	Number bonds to 10	78
☐	☐	Adding numbers to 10	80
☐	☐	Subtracting numbers to 10	82
☐	☐	Number bonds to 20	84
☐	☐	Adding numbers to 20	86
☐	☐	Subtracting numbers to 20	88
☐	☐	Problem solving	90

Multiplication and division

Practised Achieved

| ☐ | ☐ | Grouping | 92 |
| ☐ | ☐ | Sharing | 94 |

Fractions

Practised Achieved

| ☐ | ☐ | Halves | 96 |
| ☐ | ☐ | Quarters | 98 |

Measurement

Practised Achieved

☐	☐	Length	100
☐	☐	Mass and weight	102
☐	☐	Capacity and volume	104
☐	☐	Time	106
☐	☐	Money	108

Geometry: Properties of shape

Practised Achieved

| ☐ | ☐ | 2D shapes | 110 |
| ☐ | ☐ | 3D shapes | 112 |

Geometry: Position and direction

Practised Achieved

| ☐ | ☐ | Turns | 114 |

Notes

SCHOLASTIC

Help children to master core English Skills

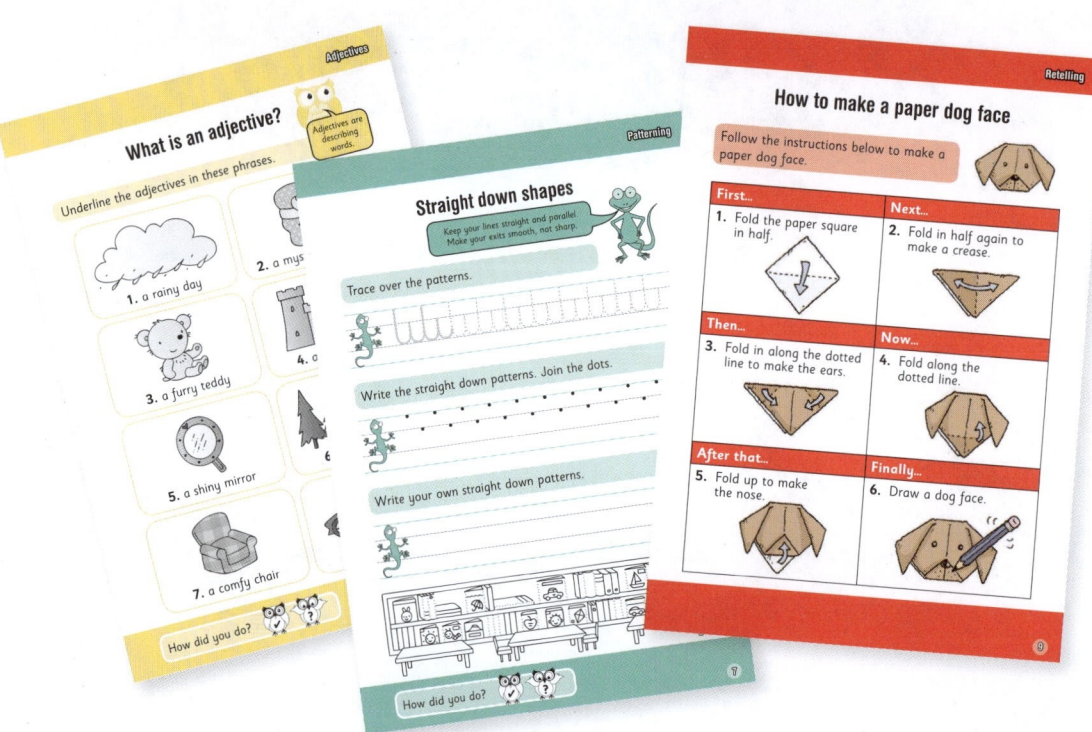

Ages 5–6 Year 1
Ages 6–7 Year 2
Ages 7–8 Year 3
Ages 9–10 Year 5

Build confidence with targeted skills practice

Master comprehension, grammar, handwriting and spelling

- Colourful practice activities to use at home or in school
- Matched to National Curriculum requirements
- Includes extra notes and tips to reinforce skills and support learning at home

Revision & Practice → **10-Minute Tests** → **National Tests** → **Catch-up & Challenge**

Available everywhere books are sold

WHSmiths amazon Waterstones

Find out more at
www.scholastic.co.uk/learn-at-home